Time Series Analysis and Forecasting by Example

Time Series Analysis and Forecasting by Example

Editor

Lavra Filipek

Time Series Analysis and Forecasting by Example

Edited by **Lavra Filipek**

Printed in 2017

ISBN: 978-1-68117-192-0

Library of Congress Control Number: 2015949136

Notice

Reasonable efforts have been made to publish reliable data and views articulated in the chapters are those of the individual contributors, and not necessarily those of the editors or publishers. Editors or publishers are not responsible for the accuracy of the information in the published chapters or consequences of their use. The publisher believes no responsibility for any damage or grievance to the persons or property arising out of the use of any materials, instructions, methods or thoughts in the book. The editors and the publisher have attempted to trace the copyright holders of all material reproduced in this publication and apologize to copyright holders if permission has not been obtained. If any copyright holder has not been acknowledged, please write to us so we may rectify.

Preface

Time series analysis comprises methods for analyzing time series data in order to extract meaningful statistics and other characteristics of the data. Time seriesforecasting is the use of a model to predict future values based on previously observed values. While regression analysis is often employed in such a way as to test theories that the current values of one or more independent time series affect the current value of another time series, this type of analysis of time series is not called "time series analysis", which focuses on comparing values of a single time series or multiple dependent time series at different points in time.

Time series data have a natural temporal ordering. This makes time series analysis distinct from cross-sectional studies, in which there is no natural ordering of the observations. Time series analysis is also distinct from spatial data analysis where the observations typically relate to geographical locations. A stochastic model for a time series will generally reflect the fact that observations close together in time will be more closely related than observations further apart. In addition, time series models will often make use of the natural one-way ordering of time so that values for a given period will be expressed as deriving in some way from past values, rather than from future values.

Time Series Analysis and Forecasting by Example emphasizes on techniques in time series analysis using various examples. The book focuses methods and techniques for time series analysis in a simplified, example-based approach. This book is concerned with forecasting methods based on the use of time-series analysis. It is primarily intended as a reference source for practitioners and researchers in forecasting, who could, for example, be statisticians, econometricians, operational researchers, management scientists or decision scientists. The book could also be used as a text for a graduate-level course in forecasting.

Table of Contents

Parameters Optimization Using Genetic Algorithms In Support Vector Regression For Sales Volume Forecasting

Fong-Ching Yuan

Department of Information Management,
Yuan Ze University, Chung-Li,
Chinese Taipei
Email: imyuan@saturn.yzu.edu.tw

1

ABSTRACT

Budgeting planning plays an important role in coordinating activities in organizations. An accurate sales volume forecasting is the key to the entire budgeting process. All of the other parts of the master budget are dependent on the sales volume forecasting in some way. If the sales volume forecasting is sloppily done, then the rest of the budgeting process is largely a waste of time. Therefore, the sales volume forecasting process is a critical one for most businesses, and also a difficult area of management. Most of researches and companies use the statistical methods, regression analysis, or sophisticated computer simulations to analyze the sales volume forecasting. Recently, various prediction Artificial Intelligent (AI) techniques have been proposed in forecasting. Support Vector Regression (SVR) has been applied successfully to solve problems in numerous fields and proved to be a better prediction model. However, the select of appropriate SVR parameters is difficult. Therefore, to improve the accuracy of SVR, a hybrid intelligent support system based on evolutionary computation to solve the difficulties involved with the parameters selection is presented in this research. Genetic Algorithms (GAs) are used to optimize free parameters of SVR. The experimental results indicate that GA-SVR can achieve better forecasting accuracy and performance than traditional SVR and artificial neural network (ANN) prediction models in sales volume forecasting.

INTRODUCTION

Sales forecasting is a self-assessment tool for a company. The managers have to keep taking the pulse of their company to know how healthy it is. A sales forecast reports, graphs and analyzes the pulse of the business. It can make the difference between just surviving and being highly successful in business. It is a vital cornerstone of a company's budget. The future direction of the company may rest on the accuracy of sales forecasting [1].

For sales forecasting to be valuable to the business, it must not be treated as an isolated exercise. Rather, it must be integrated into all facets of the organization. Thus, all enterprises are working on the exploitation of prediction methods, which decide the success and failure of an enterprise [2,3].

Business forecasting has consistently been a critical organizational capability for both strategic and tactical business planning [4]. Thus, how to improve the quality of forecasts is still an outstanding question [5]. For data containing trend or/and seasonal patterns, failure to account for these patterns may result in poor forecasts. Over the last few decades when dealing with the problems of sales forecasting, traditional time series forecasting methods, such as exponential smoothing, moving average, Box Jenkins ARIMA, and multivariate regressions etc., have been proposed and widely used in practice to account for these patterns, but it always doesn't work when the market fluctuates frequently and at random[6,7]. Therefore, Research on novel business forecasting techniques have evoked researchers from various disciplines such as computational artificial intelligence.

An artificial neural network (ANN) is a new contender in forecasting sophisticated trend and seasonal data. Artificial intelligent models have more flexibility and can be used to estimate the non-linear relationship, without the limits of traditional time series models [8]. Therefore, more and more researchers tend to use AI forecasting models to deal with forecasting problems. Artificial neural network (ANN) has strong parallel processing and fault tolerant ability. However, the practicability of ANN is affected due to several weaknesses, such as over-fitting, slow convergence velocity and relapsing into local extremum easily [9].

Support Vector Machines (SVM), a more recent learning algorithm that has been developed from statistical learning theory [10,11], has a very strong mathematical foundation and has been shown to exhibit excellent performance in time series forecasting [7,12-14] and in classification [15,16]. SVM is a new machine learning method based on the statistical learning theory, which solves the problem of over-fitting, local optimal solution and low convergence rate existed in ANN and has excellent generalization ability in the situation of small sample. When SVM is used in regression, it is called support vector regression (SVR). However, the select of appropriate SVR parameters is difficult. A highly effective model can be built after the parameters of SVR are carefully determined [17].

Whereas GA has strong global search capability[18], support vector regression optimized by genetic algorithm (GA-SVR) is proposed to forecast the sales volume, among which GA is used to determine training parameters of support vector regression [19,20]. The GA proposed by Holland [21] is derivative-free stochastic optimization method based on the concepts of natural selection and evolutionary processes. The GA also encodes each point in a parameter space into a binary bit string called a chromosome. Major components of this algorithm include encoding schemes, fitness evaluation, parent selection, crossover, and mutation operators. GASVR has been used in many fields and proved be a very effective method [11,19,22], but not used in sales volume forecasting. Therefore, in this research, the hybrid improved intelligent models, GA-SVR, will be discussed for forecasting monthly sales volume of car industry and compared with ANN and other traditional models.

The rest of the paper is organized as follows. Section 2 describes the theory of support vector regression. Section 3 presents the experiment design. The data of sales volume of a car manufacturer in Taiwan is used as a case study to test the reliability and accuracy of the proposed model. Section 4 contains experimental results and analysis. Finally, Section 5 concludes the paper.

THEORY OF SUPPORT VECTOR REGRESSION

The basic concept of SVR is that nonlinearly the original dataset x_i is mapped into a high-dimensional feature space. Given data set

$\{(x_1,y_1),\ldots,(x_1,y_1)\}$ where x_i is the input vector, y_i is the associated output value of x_i. The SVR regression function is:

$$f(x) = w \times \Phi(x) + b$$

(1)

where $\phi(x)$ denotes the non-linear mapping function, w is the weight vector and b is the bias term. The goal of SVR is to find a function f(x) that has at most ε deviation from the targets y_i for all the training data and, at the same time, is as flat as possible. In SVR, ε-insensitive loss function is introduced to ensure the sparsity of support vector, which is defined as:

$$L_\varepsilon(y_i, f(x_i))$$
$$= \begin{cases} |y_i - f(x_i)| - \varepsilon, & |y_i - f(x_i)| \geq \varepsilon \\ 0 & \text{otherwise} \end{cases}$$

(2)

where the loss equals zero if the error of forecasting values is less than ε, otherwise the loss equals value larger than ε.

As with the classification problem, non-negative slack variables, ξ_i and ξ^*, can be introduced to represent the distance from actual values to the corresponding boundary values of the ε-tube. Then, the constrained form can be formulated as follows:

$$\min \frac{1}{2}\|w\|^2 + C\left(\xi_i + \xi_i^*\right)$$

(3)

Subject to

$$y_i - \left[w \times \Phi(x)\right] - b \leq \varepsilon + \xi_i$$
$$\left[w \times \Phi(x)\right] + b - yi \leq \varepsilon + \xi_i^*$$
$$\xi_i, \xi_i^* \geq 0$$

where C denoted a cost function measuring the empirical risk.

Finally, the constrained optimization problem is solved using the following Lagrange form:

Max

$$\sum_{i=1}^{l} y_i \left(\alpha_i - \alpha_i^* \right) - \varepsilon \sum_{i=1}^{l} \left(\alpha_i + \alpha_i^* \right)$$

$$-\frac{1}{2}\sum_{i=1}^{l}\sum_{j=1}^{l}\left(\alpha_i - \alpha_i^*\right)\left(\alpha_j - \alpha_{ii}^*\right)K\left(x_i, x_j\right) \tag{4}$$

Subject to

$$\sum_{i=1}^{l}\left(\alpha_i - \alpha_i^*\right) = 0 \quad \alpha_i, \alpha_i^* \in [0, C]$$

where α_i and α_i^* are Lagrange multipliers. $K(x_1, x_j) = \varphi(x_i)\,\varphi(x_j)$ is a so-called kernel function. By using a kernel function, it is possible to compute the SVR without explicitly mapping in the feature space. The condition for choosing kernel functions should conform to Mercer's condition, which allows the kernel substitutions to represent do products in some Hilbert space. SVM constructed by Gaussian radial basis function (RBF) $\left(K\left(x_i, x_j\right) = \exp\left(-\|x_i - x_j\| / 2\sigma^2\right)\right)$ has excellent nonlinear forecasting performance. Thus, in this work, RBF is used in the SVR.

Equation (1) can now be rewritten as follows:

$$w = \sum_{i=1}^{l} \left(\alpha_i - \alpha_i^* \right) x_i, \text{ and therefore}$$

$$f(x) = \sum \left(\alpha_i - \alpha_i^* \right) K\left(x_i, x_j\right) + b \tag{5}$$

The training parameters C, σ and ε greatly affect the forecasting performance of SVR. However, the appropriate select of these SVR parameters is very difficult. In this research, GA is used to optimize the training parameters. GA has strong global search capability, which can get optimal solution in short time [20]. So GA is used to search for better combinations of the parameters in SVR. After a series of iterative computations, GA can obtain the optimal solution. The methods and process of optimizing the SVR parameters with genetic algorithm is described as follows:

Initial Value of SVR Parameters

In our proposed novel GA-SVR model, the training parameters C, σ and ε of SVR are dynamically optimized by implementing the evolutionary process with a randomly generated initial population of chromosomes, and the SVR model then performs the prediction task using these optimal values. Our approach simultaneously determines the appropriate type of kernel function and optimal kernel parameter values for optimizing the SVR model. The process of optimizing the SVR parameters with genetic algorithm is shown in **Figure 1**.

Genetic Operations

Generally, genetic algorithm uses selection, crossover and mutation operation to generate the offspring of the existing population as described as follows:

"Selection" operator: Selection is performed to select excellent chromosomes to reproduce. Based on fitness function, chromosomes with higher fitness values are more likely to yield offspring in the next generation by means of the roulette wheel or tournament method to decide whether or not a chromosome can survive into the next generation. The chromosomes that survive into the next generation are then

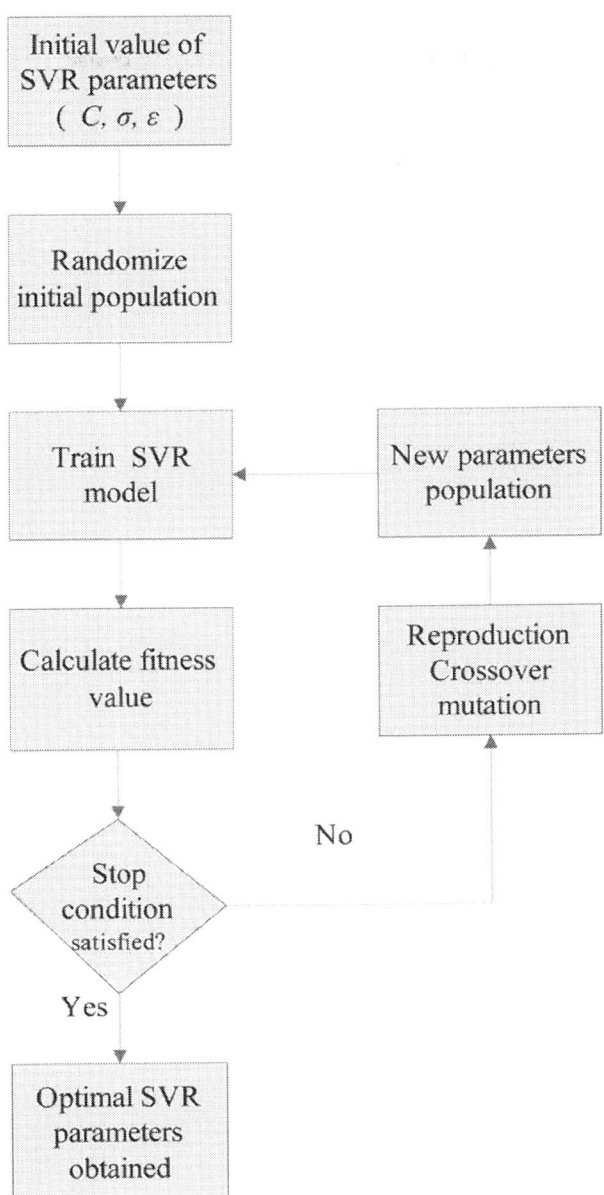

Figure 1: The process of SVR parameters optimized by genetic algorithm.

placed in a mating pool for the crossover and mutation operations. Once a pair of chromosomes has been selected for crossover, one or more randomly selected positions are assigned into the to-be-crossed chromosomes. The newly-crossed chromosomes then combine with the rest of the chromosomes to generate a new population. Suppose there are m individuals, we select [m/2] individuals but erase the others, the ones we selected are "more fitness" that means their profits are greater.

"Crossover" operator: Crossover is performed randomly to exchange genes between two chromosomes. Suppose $S_1 = \{s_{11}, s_{12}, \ldots, s_{1n}\}$, $S_2\{s_{21}, s_{22}, \ldots, s_{2n}\}$, are two chromosomes, select a random integer number $0 \leq r \leq n$, S_3, S_4 are offspring of crossover (S_1, S_2),

$$S_3 = \left\{ s_i \middle| \text{ if } i \leq r, s_i \in S_1, \text{ else } s_i \in S_2 \right\},$$

$$S_4 = \left\{ s_i \middle| \text{ if } i \leq r, s_i \in S_2, \text{ else } s_i \in S_1 \right\}.$$

"Mutation" operator: The mutation operation follows the crossover to determine whether or not a chromosome should mutate to the next generation. Suppose a chromosome $S_1 = \{s_{11}, s_{12}, \cdots, s_{1n}\}$, select a random integer number $0 \leq r \leq n$, S_3 is a mutation of S_1, $S_3 = \{s_i \mid \text{if } i \neq r$, then $s_i = s_{1i}$, else $s_i = \text{random}(s_{1i})\}$.

Offspring replaces the old population and forms a new population in the next generation by the three operations, the evolutionary process proceeds until stop conditions are satisfied.

Calculation of the Fitness Value

A fitness function assessing the performance for each chromosome must be designed before searching for the optimal values of the SVR parameters. Several measurement indicators have been proposed and employed to evaluate the prediction accuracy of models such as MAPE and RMSE in time-series prediction problems. To compare the results achieved by the present model, this research adopts Mean Absolute Percentage Error (MAPE) to evaluate the performance.

$$\text{MAPE} = \frac{\sum_{t=1}^{n} \left| (A_t - F_t)/A_t \right|}{n} \tag{6}$$

where A_t is the actual values for period t, F_t the expected value for period t and n is the number of training samples. The smaller the values of MAPE, the better the forecasting models will be. The smaller values mean that the calculating results are closer to the historic actual data.

EXPERIMENT DESIGN

In this research, the monthly sales volume of trucks and small cars of a car manufacturer in Taiwan, and other input variables, such as stock index, jobless rate, GDP per person, CPI, CCI, US dollars, Yen, Euro, and average gasoline price, were collected during the period from 2003~2009.

For comparison purpose, several commonly used forecasting models, such as Least-Mean Square Algorithm (LMS), Artificial Neural Networks (ANNs), and Support Vector Regression (SVR), are also applied.

In the experiments, the models are trained using training data, and are applied to testing data. Thus, the models are trained with input data from the year 2003 and output data (forecasted monthly sales volume) from 2004. Then the data from the year 2004 were entered as testing data in order to forecast the monthly sales volume from 2005. For later years the data from all the previous years were used in the training phase. In a subsequent cross-section analysis the mean absolute percentage error (MAPE) is used to evaluate the forecasting accuracy.

EXPERIMENTAL RESULTS AND ANALYSIS

After using the above-mentioned data and implementing the computational procedure, the monthly prediction results of forecasting models for trucks based on previous monthly sales volume and other factors

Table 1: Comparison of the prediction results for trucks from each model for 12 months from 2005-2009

Year	Model	Jan	Feb	Mar	Apr	May	Jun	Jul	Aug	Sep	Oct	Nov	Dec
2008	BPN	1675	636	686	584	529	548	857	882	1582	1649	2081	2615
	SVR	1506	1298	1746	1591	1409	1077	702	318	353	1016	539	234
	GA-SVR	785	397	444	404	341	127	115	106	209	546	548	371
	AC-TUAL	664	414	403	512	503	702	848	749	676	1172	1128	1326
2009	LMS	-278	-486	-382	-386	-503	-432	249	29	III	1202	1204	1216
	BPN	405	380	374	382	358	351	71	71	72	154	317	253
	SVR	1478	1207	1203	1376	1407	1843	1804	1664	1448	2019	1896	2169
	GA-SVR	401	401	551	551	551	551	523	530	526	600	723	945
Years		Jan	Feb	Mar	Apr	May	Jun	Jul	Aug	Sep	Oct	Nov	Dec
2005	AC-TUAL	5597	1226	2951	2196	1927	2262	2564	1701	2019	1502	1520	1175
	LMS	4164	2187	2137	1102	1661	799	-853	-523	-534	-2539	-2987	-4361
	BPN	3380	1205	1144	942	1304	1219	1053	974	1064	1846	2287	864
	SVR	7703	2128	3843	2735	2111	2946	4537	3270	3492	3179	3413	3649
	GA-SVR	2117	1895	1783	1480	1830	1943	1831	1562	1529	1417	1437	1169
	AC-TUAL	4447	1339	1470	1606	1462	1078	911	721	719	764	852	408

Year	Model												
2006	LMS	4588	2468	2839	3375	3779	2940	2270	1601	2415	2882	3582	3672
	BPN	4698	3390	3242	4707	4855	2683	2110	1595	4319	4105	3887	2527
	SVR	4734	2030	2496	2409	2168	1566	1397	940	1928	2224	2465	1885
	GA-SVR	2379	1178	1611	1288	1062	1118	1005	883	932	866	1002	944
	AC-TUAL	2758	1049	1102	754	592	601	724	498	626	944	726	1503
2007	LMS	1523	1317	603	734	1315	2499	2721	1793	1744	2557	743	80
	BPN	2501	528	486	626	381	390	355	731	1196	1760	1651	1185
	SVR	3747	1364	1205	888	1010	1218	772	637	676	1644	1899	3962
	GA-SVR	1212	1065	951	896	896	823	772	633	672	606	548	527
	AC-TUAL	626	219	633	495	386	388	268	105	277	835	401	187
	LMS	727	530	1002	1297	610	479	−59	−177	−692	−615	−1767	−3180

during 2003-2009 are summarized in **Table 1** and Figures 2-6, respectively. Accordingly, **Table 2" target="_self"> Table 2** and **Figure 7** report the MAPE results. All experimental results indicate that GA-SVR has more excellent performance than other models in forecasting monthly sales volume.

Considering the adaptability of GA-SVR, we also use four models above to predict the small car data set. As shown in **Table 3** and **Figure 8**, the experimental results indicate that GA-SVR model can achieve better forecasting accuracy and performance than other models.

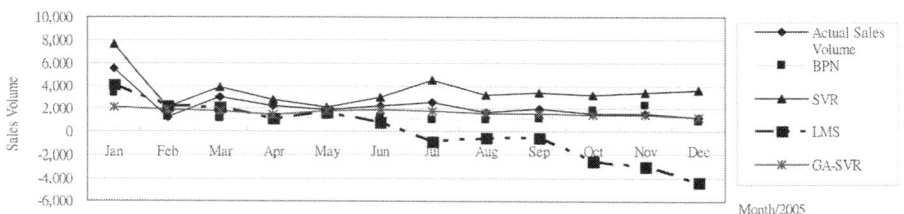

Figure 2: The 2005 monthly prediction results for trucks based on 2004 testing data.

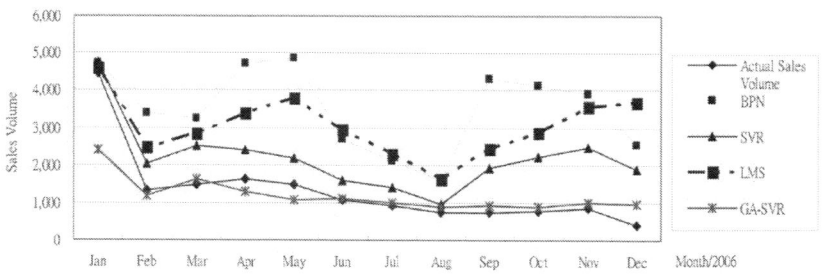

Figure 3: The 2006 monthly prediction results for trucks based on 2005 testing data.

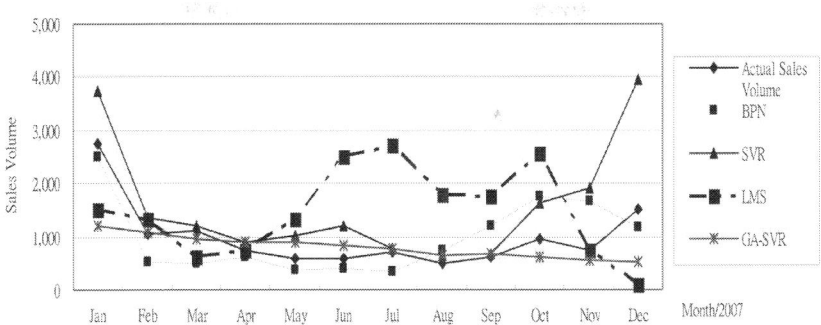

Figure 4: The 2007 monthly prediction results for trucks based on 2006 testing data.

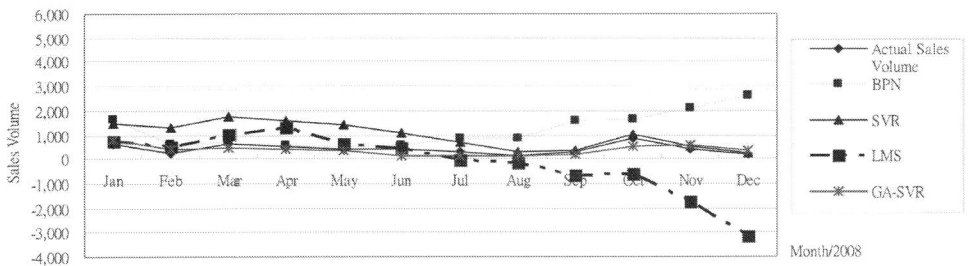

Figure 5: The 2008 monthly prediction results for trucks based on 2007 testing data.

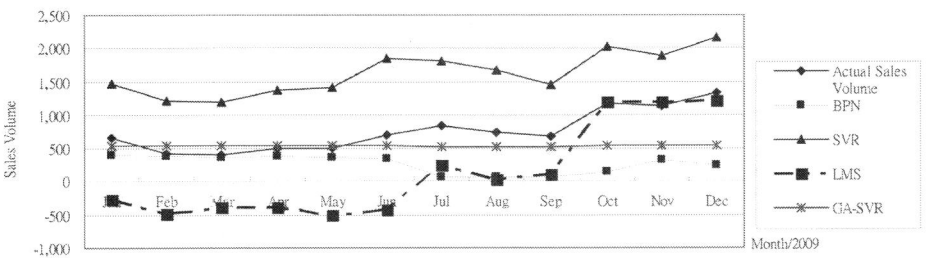

Figure 6: The 2009 monthly prediction results for trucks based on 2008 testing data.

Table 2: Comparison of the MAPE results for trucks

Al Lzorithm Year	LMS	BPN	SVR	GA-SVR
2005	140.58%	40.58%	74.55%	23.40%
2006	210.55%	243.32%	105.48%	28.64%
2007	128.21%	52.20%	59.00%	28.75%
2008	309.58%	309.09%	214.04%	40.59%
2009	113.25%	55.80%	131.38%	26.76%
Average	180.44%	140.20%	116.89%	29.63%

Table 3: Comparison of the MAPE results for small cars

Al u·orithm Year	LMS	BPN	SVR	GA-SVR
2005	51.16%	34.70%	33.40%	36.01%
2006	223.81%	190.24%	67.48%	20.08%
2007	48.45%	29.84%	51.22%	18.08%
2008	266.97%	276.61%	129.50%	46.03%
2009	83.90%	48.66%	152.59%	30.20%
Average	134.86%	116.01%	86.84%	30.08%

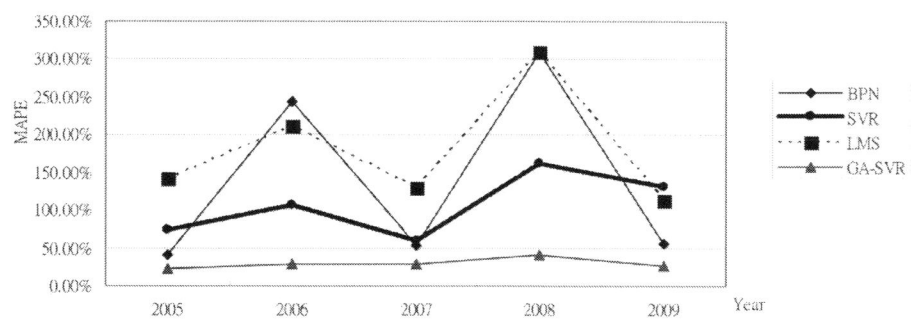

Figure 7: Comparison of the MAPE results for trucks.

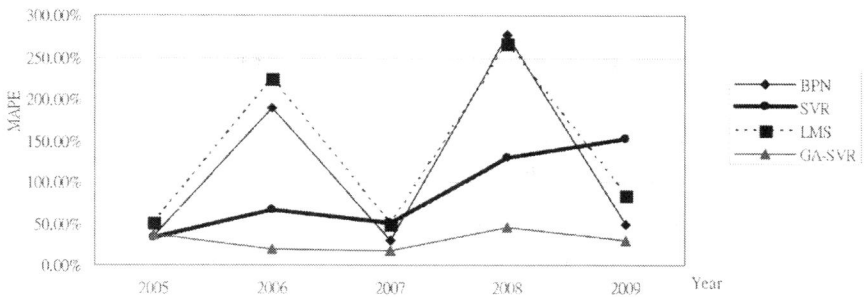

Figure 8: Comparison of the MAPE results for small cars.

CONCLUSION

The sales forecasting process is a critical one for most business. GA-SVR is applied to forecast the car sales volume in this study. In GA-SVR model, GA is used to select optimal parameters of SVR, among which the MAPE method is determined to evaluate fitness. The monthly sales volume of a car manufacturer in Taiwan from 2003 to 2009 is used as our research data. The experimental results show that GA-SVR can achieve a greater forecasting accuracy than artificial neural network and other traditional models. Even the WTO impact to Taiwan car industry in 2006 and the unexpected financial tsunami in 2008, GA-SVR forecasting capability still outperforms than other methods.

ACKNOWLEDGEMENTS

The author would like to thank the National Science Council, Taiwan, ROC for financially supporting this research under Contract Nos. NSC101-2410-H-155-004.

REFERENCES

1. R. H. Garrison and E. W. Noreen, "Managerial Accounting 10/e," McGraw-Hill, New York, 2003.
2. J. Stack, "A Passion for Forecasting," Springfield Manufacturing, Inc., Springfield, 1997, pp. 37-38.
3. C. D. Lewis, "Industrial and Business Forecasting Methods," Butterworths, London, 1982.
4. R. Fildes, R. Hastings, "The Organization and Improvement of Market Forecasting," Journal of Operation Research Society, Vol. 45, No. 1, 1994, pp. 1-16.
5. C. W. J. Granger, "Can We Improve the Perceived Quality of Economic Forecasts?" Journal Applied Econometric, Vol. 11, No. 5, 1996, pp. 455-473.
6. M. Lawrence and M. O'Connor, "Sales Forecasting Updates: How Good Are They in Practice?" International Journal of Forecasting, Vol. 16, No. 3, 2000, pp. 369- 382.doi:10.1016/S0169-2070(00)00059-5.
7. Y. K. Bao, Y. S. Lu and J. L. Zhang, "Forecasting Stock Price by SVMs Regression," Lecture Notes in Artificial Intelligence, Vol. 3192, 2004, pp. 295-303.
8. R. J. Kuo and K. C. Xue, "A Decision Support System for Sales Forecasting through Fuzzy Neural Networks with Asymmetric Fuzzy Weights," Decisions Support Systems, Vol. 24, No. 2, 1998, pp. 105-126. doi:10.1016/S0167-9236(98)00067-0.
9. R. Capparuccia, R. De. Leone, E. Marchitto, "Integrating Support Vector Machines and Neural Networks," Neural Networks, Vol. 20, No. 5, 2007, pp. 590-597.doi:10.1016/j.neunet.2006.12.003.
10. V. N. Vapnik, "The Nature of Statistical Learning Theory," Springer, New York, 1995.
11. V. Vapnik, S. Golowich and A. Smola, "Support Vector Method for Function Approximation, Regression Estimation and Signal Processing," In: M. Mozer, M. Jordan and T. Petsche, Eds., Advance in Neural Information Processing System, MIT Press, Cambridge, 1997, pp.281-287.
12. K. R. Muller, A. J. Smola, G. Ratsch, B. Scholkopf, J. Kohlmorgen and V. Vapnik, "Prediction Time Series with Support Vector Machines," Lecture Notes in Computer Science, Vol. 1327, 1997, pp. 999-1004. doi:10.1007/BFb0020283.
13. E. H. T. Francis and L. Cao, "Application of Support Vector Machines in Financial Time Series Forecasting," International Journal of Management Science, Vol. 29, No. 4, 2001, pp. 309-317.
14. B. Yukun, R. Zhang and S. F. Crone, "Fuzzy Support Vector Machines Regression for Business Forecasting: An Application," Fuzzy Systems and Knowledge Discovery, Vol. 4223, 2006, pp. 1313-1317.
15. L. Yu, S. Wang and J. Cao, "A Modified Least Squares Support Vector Machine Classifier with Application to Credit Risk Analysis," International Journal of Information Technology & Decision Making, Vol. 8, No. 4, 2009, pp. 697-710. doi:10.1142/S0219622009003600.

16. C. H. Zheng, G. W. Zheng and L. C. Jiao, "Heuristic Genetic Algorithm-Based Support Vector Classifier for Recognition of Remote Sensing Images," In: Advance in Neural Networks, Lecture Notes in Computer Science, Springer-Verlag, New York, Vol. 3173, 2004, pp. 629- 635.

17. K. Duan, S. S. Keerthi, A. N. Poo, "Evaluation of Simple Performance Measures for Tuning SVM Hyperparameters," Neurocomputing, Vol. 51, No. 1-4, 2003, pp. 41-59.doi:10.1016/S0925-2312(02)00601-X

18. M. Kaya, "MOGAMOD: Multi-Objective Genetic Algorithm for Motif Discovery," Expert Systems with Applications, Vol. 36, No. 2, 2007, pp. 1039-1047. doi:10.1016/j.eswa.2007.11.008

19. X. G. Chen, "Railway Passenger Volume Forecasting Based on Support Machine and Genetic Algorithm," 2009 ETP International Conference on Future Computer and Communication, 6-7 June 2009, Wuhan, pp. 282-284. doi:10.1109/FCC.2009.81

20. C. H. Wu, G. H. Tzeng and R. H. Lin, "A Novel Hybrid Genetic Algorithm for Kernel Function and Parameter Optimization in Support Vector Regression," Expert Systems with Applications, Vol. 36, No. 3, 2009, pp. 4725- 4735. doi:10.1016/j.eswa.2008.06.046

21. J. H. Holland, "Adaptation in Natural and Artificial Systems," University of Michigan, MIT Press, Cambridge, 1975.

22. P. F. Pai and W. C. Hong, "Forecasting Regional Electric Load Based on Recurrent Support Vector Machines with Genetic Algorithms," Electric Power Systems Research, Vol. 74, No. 3, 2005, pp. 417-425. doi:10.1016/j.epsr.2005.01.006.

CITATION

F. Yuan, "Parameters Optimization Using Genetic Algorithms in Support Vector Regression for Sales Volume Forecasting," Applied Mathematics, Vol. 3 No. 10A, 2012, pp. 1480-1486. Doi: 10.4236/am.2012.330207.

Optimal Redundancy Allocation in Hierarchical Series—Parallel Systems Using Mixed Integer Programming

Mohsen Ziaee
Department of Industrial Engineering,
University of Bojnord, Bojnord, Iran

ABSTRACT

Reliability optimization plays an important role in design, operation and management of the industrial systems. System reliability can be easily enhanced by improving the reliability of unreliable components and/or by using redundant configuration with subsystems/components in parallel. Redundancy Allocation Problem (RAP) was studied in this research. A mixed integer programming model was proposed to solve the problem, which considers simultaneously two objectives under several resource constraints. The model is only for the hierarchical series-parallel systems in which the elements of any subset of subsystems or components are connected in series or parallel and constitute a larger subsystem or total system. At the end of the study, the performance of the proposed approach was evaluated by a numerical example.

INTRODUCTION

The reliability optimization plays an important role in design, operation and management of the industrial systems. The reliability of a system can be easily enhanced by improving the reliability of unreliable components and/or by using redundant configuration with subsystems/

components in parallel. Improving component reliability has been generally preferred over adding redundancy in industry, because the redundancy is difficult to add to the real systems due to the technical limitations such as weight, volume, and cost. However, recently developed advanced technologies such as semiconductor integrated circuits and nanotechnology, have revived the importance of the redundancy strategy [1].

A well-known and complex reliability optimization problem is the redundancy apportionment problem for the series-parallel systems which can be defined as the problem of the selection of the optimal combination of component type and redundancy level for each subsystem in order to meet various objectives under given constraints on the overall system [2]. This problem was studied in this research. The objectives considered were the maximization of the system reliability, and the minimization of the system cost. The study was looking for the number and the type of the redundant components which optimize the objective function under several different constraints such as the overall system weight and total number of the components used in all redundancies.

The literature abounds with numerous and very different techniques for solving the optimal redundancy allocation problem with various objectives and different resource constraints. Ha and Kuo [1] presented a branchand-bound approach to solve the redundancy allocation problem (RAP) formulated as a non-convex integer nonlinear programming model. Their computational experiments demonstrated that the method was superior to the other existing exact algorithms for RAP in terms of computation time. A combined approach was presented by Nourelfath and Dutuit [3] to solve the redundancy optimization problem for multi-state systems under repair policies. Azaron et al. [4] dealt with the reliability function of a class of time-dependent systems with stand by redundancy. You and Chen [5] proposed a heuristic algorithm based on a multi-start search procedure for solving a series-parallel RAP with separable linear constraints. An Ant Colony Optimization (ACO) algorithm was proposed by Liang and Smith [6] for the RAP. Shelokar et al. [7] applied an ACO algorithm for single and multi-objective reliability optimization problems. Nahas and Nourelfath [8] examined applying an ant system to reliability optimization of a series system with multiple-choice and budget constraints. An ACO algorithm with a multi-objective formulation was developed by Zhao et al. [2] in order to solve the redundancy apportionment problem of seriesparallel k-out-of-n G: subsystems (denoted by ACSRAP). Mahapatra and Roy [9] dealt with the reliability optimization problem with several mutually conflicting objectives, which were the minimization of the system cost and the maximization of the system reliability, and proposed a fuzzy multi-objective optimization method for the series and complex system reliability. Ouzineb et al. [10] presented a heuristic approach based on a combination of space partitioning, genetic algorithm and tabu search to solve the redundancy allocation problem for series—parallel binary-state systems. The design goal of the RAP was to select the optimal combination of elements and redundancy levels to maximize system reliability subject to the system budget and to the system weight. After dividing the search space into a set of disjoint subsets, this approach uses GA to select the subspaces, and applies TS to each selected subspace. A heterogeneous redundancy optimization model based on genetic algorithm was proposed by Li et al. [11] for multistate series—parallel systems subject to common cause failures, in order to provide a desired level of reliability with minimum cost. Recently, Sharma et al. [12] investigated the heterogeneous RAP in multi-state series parallel reliability structures with the objective of the minimization of the total cost of system design satisfying given reliability constraint and consumer load demand. The demand distribution was presented as a piecewise cumulative load curve and each subsystem was allowed to consist of parallel redundant components of not more than three types. They proposed an ACO algorithm to solve the problem. There are many other researches on this topic in the literature (see, for example [13-16]). This study dealt with the RAP for hierarchical series—parallel systems and a new approach for mathematically modeling the problem was presented.

ASSUMPTIONS AND NOTATIONS

The assumptions considered in this study were as follows:

1. The model was only for the hierarchical series—parallel systems. A reliability system is called a Hierarchical Series Parallel system (HSP) if the system can be viewed as a set of subsystems arranged in a series parallel; each subsystem has a similar configuration; subsystems of each subsystem have a similar configuration and so on.
2. It was assumed that in a parallel configuration, at least one active component is required for the function; and in a series configuration, all components have to be active for the function.
3. The constrained resources considered in this problem were the repair times (in man-months, for example) and total number of components used in all redundancies.
4. The overall system weight (including all redundancies) could not exceed its upper limit.
5. Although in many real-world optimization problems, the financial budget is the most important constrained resource, in the reliability optimization problems, it is usually of less importance than technical constraints such as lower limit of whole system reliability; and therefore, in the problem studied in this paper, it was not considered as a constraint of the model and inadequacy of the system reliability coming from the budget limitation would not occur. Instead, the cost minimization was taken into account as an objective.
6. The maximization of the overall system reliability was also considered as an objective; moreover, it was assumed that the overall system reliability can not be less than its lower limit.
7. For each subsystem (component or system), there were several choices of subsystems (components or systems) with different reliability and resource requirements which could be used as redundancies.

The following notations were used in the proposed method.

Indices

i: If the total system is decomposed into several series/ parallel subsystems again and again until indecomposable subsystems (i.e. components) are reached, then a subsystem is at level i if it is result of decomposing the total system to series/parallel subsystems $(n - i)$ times (if the total system can be decomposed up to n levels). We have $i = 0, ..., n$; i $= n$ for the total system, and $i = 0$ for the components at the last level).

j: denotes jth subsystem of a given decomposition level (say i), $j = 1$, 2, ..., m_i; i.e. at level i, there are m_i subsystems. Note that a subsystem may be a component or a set of components configured as a series/ parallel system.

k: denotes kth redundant subsystem for a given subsystem.

Parameters

k_j^i : Number of redundant subsystems for subsystem j at level i. These redundancies have different characteristics and performances, i.e. different reliabilities, procurement costs, repair costs, ..., and any arbitrary subset of them can be used as redundant subsystems for that subsystem; so $k = 1, 2, ..., k_j^i$.

Cp_{jk}^i : Procurement cost of redundant subsystem k for subsystem j at level i.

Cr_{jk}^i : Repair cost of redundant subsystem k for subsystem j at level i.

R_{jk} : Reliability of redundant subsystem k for subsystem j at level i.

λ_{jk}^i : Failure rate of redundant subsystem k for subsystem j at level i.

t_{jk}^i : Repair time of redundant subsystem k for subsystem j at level i.

w^i_{jk} : Weight of redundant subsystem k for subsystem j at level i.

N^i_{jk} : Number of components used in redundant subsystem k for subsystem j at level i. Therefore, if the subsystem is a component, then its N^i_{jk} is equal to 1.

RS^i_j : Reliability of subsystem j at level i.

λS^0_j : Failure rate of subsystem j at level 0 (which is a component).

t^0_j : Repair time of subsystem j at level 0 (i.e. a component).

WS^0_j : Weight of subsystem j at level 0 (i.e. a component).

L: A scaling parameter.

T: Mission time.

Tr: Total available repair time (for example manmonths).

R: A lower limit on the overall system (including its redundant subsystems) reliability.

W: Maximum weight of the overall system (including its redundant subsystems).

A: An upper limit on the number of all redundant components used in the system.

Variables

X^i_{jk} : Binary variable taking value 1 if redundant subsystem k for subsystem j at level i is used for improving the system reliability and 0 otherwise.

TRS_j^i (TRS_j^{ia} / TRS_j^{ib}): Total reliability of subsystem j at level i (including its redundant subsystems). The letter a (letter b) denotes that the subsystem is constituted of several subsystems connected in series (parallel).

MIXED INTEGER PROGRAMMING MODEL

In this section, a mixed integer programming formulation is presented to solve the problem.

$$-\sum_{i=0}^{n}\sum_{j=1}^{m_i}\sum_{k=1}^{k_j^i}\left(\left(Cp_{jk}^i + T \cdot \lambda_{jk}^i \cdot Cr_{jk}^i\right) \cdot X_{jk}^i\right)$$

Subject to:

$$TRS_j^{ia} + \left(1 - \prod_{l=1}^{m_{j\text{-}1}} TRS_l^{i\text{-}1}\right) \cdot \prod_{k=1}^{k_j^i}\left(1 - R_{jk}^i \cdot X_{jk}^i\right) = 1$$

$\forall j; \ 1 \leq i \leq n$ \hfill (1)

$$TRS_j^{ib} + \left\{1 - \left[1 - \prod_{l=1}^{m_{j\text{-}1}}\left(1 - TRS_l^{i\text{-}1}\right)\right]\right\}\prod_{k=1}^{k_j^i}\left(1 - R_{jk}^i \cdot X_{jk}^i\right)$$

$= 1 \qquad \forall j; \ 1 \leq i \leq n$ \hfill (2)

$$\sum_{j=1}^{m_0}\left(\lambda S_j^0 \cdot T \cdot t_j^0\right) + \sum_{i=0}^{n}\sum_{j=1}^{m_i}\sum_{k=1}^{k_j^i}\left(T \cdot \lambda_{jk}^i \cdot t_{jk}^i \cdot X_{jk}^i\right) \leq Tr$$

\hfill (3)

$$TRS_1^{na} + TRS_1^{nb} \geq R$$

\hfill (4)

$$\sum_{j=1}^{m_0}\left(WS_j^0\right) + \sum_{i=0}^{n}\sum_{j=1}^{m_i}\sum_{k=1}^{k_j^i}\left(W_{jk}^i \cdot X_{jk}^i\right) \leq W$$

\hfill (5)

$$\sum_{i=0}^{n}\sum_{j=1}^{m_i}\sum_{k=1}^{k_j^i}\left(N_{jk}^i \cdot X_{jk}^i\right) \le A$$

$$(6)$$

$$TRS_j^i, TRS_j^{ia}, TRS_j^{ib} \ge 0 \quad \forall i, j$$

$$X_{jk}^i \in \{0,1\} \quad \forall i, j, k$$

In the above model, TRS_1^{nb} TRS_1^{na} is the reliability of the overall system if its subsystems are connected in series (parallel). Constraint sets (1) and (2) are recursive equations for calculating the reliability of any subsystem. Constraint (3) ensures that total required repair time does not exceed total available repair time. Constraint (4) guarantees that the overall system reliability is not less than its lower limit. Constraint sets (5) and (6) ensure that the overall system weight and the total number of components used in redundancies do not exceed their upper limits.

NUMERICAL EXAMPLE

The proposed method was applied to solve a test problem. Figure 1 shows this problem of the hierarchical seriesparallel system configuration. The parameter values for the problem are listed in Table 1.

Other input data are L = 2, A = 10, T = 0.5, Tr = 0.98, R = 30, W = 8, $k_j^i = 2 \forall i, j$. The results obtained from solving the problem were as follows:

The objective function value was −590.0054,

$$TRS_1^{2b} = 0.997312, \quad TRS_1^{1a} = 0.720,$$

$$TRS_1^{1b} = 0.9904, \quad TRS_1^0 = 0.800,$$

$$TRS_3^0 = 0.880 \,, \quad TRS_4^0 = 0.920 \,,$$

$$TRS_2^0 = 0.900 \,,$$

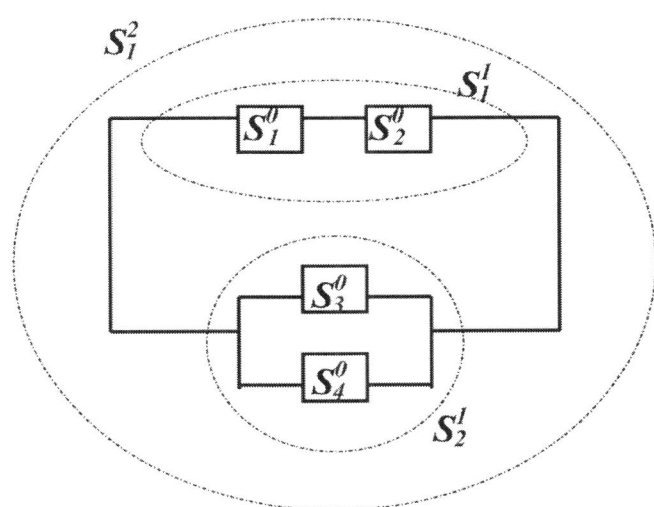

Figure 1: A hierarchical series-parallel system (S_j^i denotes subsystem j at level i).

CONCLUSIONS

In this paper, the RAP for a hierarchical series-parallel system under several resource constraints was studied. The following two objectives were considered: the maximization of the system reliability, and the minimization of the system cost. A new approach for mathematically modeling the problem was presented. The implementation of the proposed approach was illustrated by a sample application on a numerical example. Further work can be performed on to adapt the approach to other objectives and to extend it to more complex systems.

Table 1: Parameter values for the test problem

	λS_j^0	WS_j^0	RS_j^i	Cp_{j1}^i	Cp_{j2}^i	Cr_{j1}^i	Cr_{j2}^i	λ_{j1}^i
S_1^0	0.003	2	2	0.8	8	7	2	3	2
S_2^0	0.004	1	2	0.9	9	8	1	2	1
S_3^0	0.002	3	3	0.7	7	6	5	4	3
S_4^0	0.003	4	4	0.6	6	10	4	5	4
S_1^1	-	-	-	0.72	8	7	2	3	3
S_2^1	-	-	-	0.88	9	8	1	2	2
S_1^2	-	-	-	0.9664	10	9	1	1	1

	λ_{j2}^i	N_{j1}^i	N_{j2}^i	R_{j1}^i	R_{j2}^i	w_{j1}^i	w_{j2}^i	t_{j1}^i	t_{j2}^i
S_1^0	3	1	1	0.8	0.7	2	2	0.003	0.002
S_2^0	2	1	1	0.9	0.8	2	3	0.004	0.003
S_3^0	4	1	1	0.7	0.6	3	4	0.003	0.002
S_4^0	5	1	1	0.6	0.5	4	3	0.002	0.001
S_1^1	4	2	1	0.7	0.6	2	1	0.006	0.005
S_2^1	3	2	3	0.8	0.7	2	2	0.008	0.007
S_1^2	2	2	2	0.9	0.8	1	2	0.009	0.008

$X_{32}^0 = 1; X_{41}^0 = 1; X_{42}^0 = 1$ All other X_{jk}^i 's were zero.

REFERENCES

1. C. Ha and W. Kuo, "Reliability Redundancy Allocation: An Improved Real-ization for Nonconvex Nonlinear Programming Problems," European Journal of Operational Research, Vol. 171, No. 1, 2006, pp. 24-38. doi:10.1016/j.ejor.2004.06.006

2. J.-H. Zhao, Z. Liu and M.-T. Dao, "Reliability Optimization Using Multi-Objective Ant Colony System Approaches," Reliability Engineering and System Safety, Vol. 92, No. 1, 2007, pp. 109-120. doi:10.1016/j.ress.2005.12.001

3. M. Nourelfath and Y. Dutuit, "A Combined Approach to Solve the Redundancy Optimization Problem for MultiState Systems under Repair Policies," Reliability Engineering and System Safety, Vol. 86, No. 3, 2004, pp. 205- 213.doi:10.1016/j.ress.2004.01.008

4. A. Azaron, H. Katagiri, M. Sakawa and M. Modarres, "Reliability Function of a Class of Time-Dependent Systems with Standby Redundancy," European Journal of Operational Research, Vol. 164, No. 2, 2005, pp. 378- 386. doi:10.1016/j.ejor.2003.10.044

5. P.-S. You and T.-C. Chen, "An Efficient Heuristic for Series-Parallel Redundant Reliability Problems," Computers & Operations Research, Vol. 32, No. 8, 2005, pp. 2117- 2127.doi:10.1016/j.cor.2004.02.003

6. Y.-C. Liang and A. E. Smith, "An Ant Colony Optimization Algorithm for the Re-dundancy Allocation Problem (RAP)," IEEE Transactions on Reliability, Vol. 53, No. 3, 2004, pp. 417-423. doi:10.1109/TR.2004.832816

7. P. S. Shelokar, V. K. Jayaraman and B. D. Kulkarni, "Ant Algorithm for Single and Multi-Objective Reliability Optimization Problems," Quality and Reliability Engineering International, Vol. 18, No. 6, 2002, pp. 497-514. doi:10.1002/qre.499

8. N. Nahas and M. Nourelfath, "Ant System for Reliability Optimization of a Series System with Multiple-Choice and Budget Constraints," Reliability Engineering and System Safety, Vol. 87, No. 1, 2005, pp. 1-12. doi:10.1016/j.ress.2004.02.007

9. G. S. Mahapatra and T. K. Roy, "Fuzzy Multi-Objective Mathematical Programming on Reliability Optimization Model," Applied Mathematics and Computation, Vol. 174, No. 1, 2006, pp. 643-659. doi:10.1016/j.amc.2005.04.105

10. M. Ouzineb, M. Nourelfath and M. Gendreau, "An Efficient Heuristic for Reliability Design Optimization Problems," Computers & Operations Research, Vol. 37, No. 2, 2010, pp. 223-235. doi:10.1016/j.cor.2009.04.011

11. C.-Y. Li , X. Chen, X.-S. Yi and J.-Y. Tao, "Heterogeneous Redundancy Optimization for Multi-State Series-Parallel Systems Subject to Common Cause Failures," Reliability Engineering & System Safety, Vol. 95, No. 3, 2010, pp. 202-207. doi:10.1016/j.ress.2009.09.011

12. V. K. Sharma, M. Agarwal and K. Sen, "Reliability Evaluation and Optimal Design in Heterogeneous Multi-State Series-Parallel Systems," Information Sciences, Vol. 181, No. 2, 2011, pp. 362-378. doi:10.1016/j.ins.2010.09.015

13. M. S. Chern, "On the Computational Complexity of Reliability Redundancy Allocation in a Series System," Operations Research Letters, Vol. 11, No. 5, 1992, pp. 309- 315.doi:10.1016/0167-6377(92)90008-Q
14. S. B. Graves, D. C. Murphy and J. L. Ringuest, "Acceptance Sampling and Reliability: The Tradeoff between Component Quality and Redundancy," Computers & Industrial Engineering, Vol. 38, No. 1, 2000, pp. 79-91. doi:10.1016/S0360-8352(00)00030-9
15. T. Nakagawa and K. Yasui, "Note on Optimal Redundant Policies for Reliability Models," Journal of Quality in Maintenance Engineering, Vol. 11, No. 1, 2005, pp. 82- 96.doi:10.1108/13552510510589398
16. K. Y. K. Ng and N. G. F. Sancho, "A Hybrid Dynamic Programming/Depth-First Search Algorithm with an Application to Redundancy Allocation," IIE Transactions, Vol. 33, No. 12, 2001, pp. 1047-1058. doi:10.1080/07408170108936895

CITATION

M. Ziaee, "Optimal Redundancy Allocation in Hierarchical Series-Parallel Systems Using Mixed Integer Programming," Applied Mathematics, Vol. 4 No. 1, 2013, pp. 79-83. doi: 10.4236/am.2013.41014.

The Application of Time Series Modelling and Monte Carlo Simulation: Forecasting Volatile Inventory Requirements

Robert Davies, Tim Coole, David Osipyw

Faculty of Design Media and Management,
Buckinghamshire New University,
High Wycombe, UK

3

ABSTRACT

During the assembly of internal combustion engines, the specific size of crankshaft shell bearing is not known until the crankshaft is fitted to the engine block. Though the build requirements for the engine are consistent, the consumption profile of the different size shell bearings can follow a highly volatile trajectory due to minor variation in the dimensions of the crankshaft and engine block. The paper assesses the suitability of time series models including ARIMA and exponential smoothing as an appropriate method to forecast future requirements. Additionally, a Monte Carlo method is applied through building a VBA simulation tool in Microsoft Excel and comparing the output to the time series forecasts.

INTRODUCTION

Inventory control is an essential element within the discipline of operations management and serves to ensure sufficient parts and raw materials are available for immediate production needs while minimising the overall stock holding at the point of production and throughout the supply chain. Methodologies including Materials Requirements Planning and Just-in-Time Manufacturing have evolved to manage the complexities of supply management supported by an extensive academic literature. Similarly extensive study into the inventory profiles for spare part and service demand has also been prevalent over the last 50 years.

This paper presents a study of an inventory process that does not fit into the standard inventory models within conventional operations management or for service and spare parts management. In high volume internal combustion engine manufacturing, the demand profile for crankshaft shell bearings follows a highly variable demand profile though there is consistent demand for the engines. The paper reviews the application of both time series analysis and a Monte Carlo simulation method to construct a robust forecast for the shell usage consumption. Section 2 presents the problem statement. Time series analysis is reviewed in Section 3. The Monte Carlo simulation method written in Microsoft Excel VBA is presented in Section 4. Forecasts generated by both the time series models and the simulation are assessed in Section 5 and concluding remarks are presented in Section 6. The analysis utilises usage data provided by a volume engine manufacturer over a 109-week production build period.

PROBLEM STATEMENT

The construction of petrol and diesel engines involve fitting a crankshaft into an engine block and attaching piston connecting rods to the crankshaft crank pins [1] . The pistons via the connecting rods turn the crankshaft during the power stroke of the engine cycle to impart motion. Shell bearings are fitted between the crank journals and crank pins and the engine block and connecting rods. During engine operation a thin film of oil under pressure separates the bearing surface from the crankshaft journals and pins. For the smooth operation of the engine, the thickness of the oil film has to be consistent for each crank journal and pin. Though the crankshaft, connecting rods and engine block are machined to exceptionally tight tolerances minor deviations in the dimensions of the machined surfaces mean that the thickness of the oil film will not be consistent across the individual crank journals and pins.

To overcome the problem, the engine designer specifies a range of tolerance bands within an overall tolerance for the nominal thickness of the shell bearing. Similar tolerance bands are specified for the machined surfaces of the engine block and crankshaft. A shell bearing whose thickness is defined within a specific tolerance band is identified by a small colour mark applied during the shell manufacturing process. The engine designer creates a "Fitting Matrix", where the combination of the tolerance bands for the engine block and crankshaft are compared against which the appropriate shell bearing can be selected. During the engine assembly, the selection process is automated. Embossed marks on the crankshaft and engine block specify the tolerance bands of each machined surface. The marks are either optically scanned or fed into a device that creates a visual display of the colours of the shells to select for assembly. Selection is rapid and does not impede the speed of engine assembly. Table 1 illustrates the set of tolerance bands and associated "Fitting Matrix" for the selection of main bearing shells for a typical engine.

Over time the usage profile for some shells thicknesses can show considerable variation against a consistent demand for the engine itself. The high variation poses a difficult procurement problem for engine assemblers that support high volume automotive production. It is difficult to determine what the usage consumption for high usage shells will be over a future time horizon as the shell bearings can be globally sourced and can require lead times of between 3 and 6 months.

Figure 1 illustrates the consumption trajectory of journal shell sizes over a 109-week sample period that has high demand profiles. The shells are identified by the colour green and yellow. The green shell consumption is highly volatile with sporadic spikes in demand. The yellow shell consumption after showing a rapid decline over.

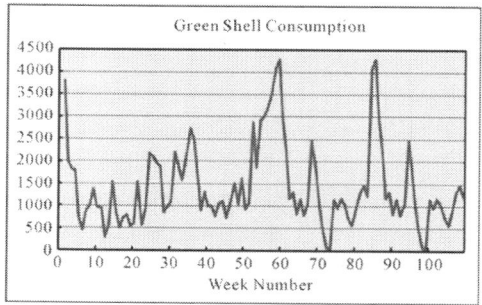 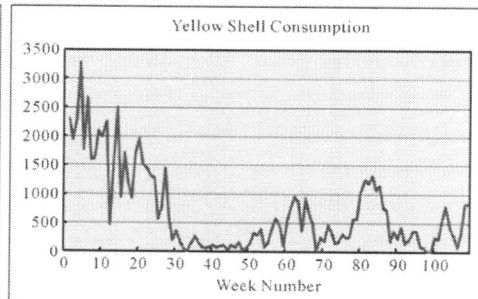

Figure 1: Time series trajectories for green and yellow shell consumption.

The first 30 weeks of the sample period settles to lower and less volatile demand than the green shell.

TIME SERIES ANALYSIS

Succinctly, a time series is a record of the observed values of a process or phenomena taken sequentially over time. The observed values are random in nature rather than deterministic where the random behaviour is more suitable to model through the laws of probability. Systems or processes that are non-deterministic in nature are referred to as stochastic where the observed values are modelled as a sequence of random variables. Formally:

A stochastic process $\{Y(t); t \in T\}$ is a collection of random variables where T is an index set for which all the random variables, $Y(t); t \in T$ are defined on the same sample space. When the index set T represents time, the stochastic process is referred to as a time series [2].

While time series analysis has extensive application, the purpose of the analysis is two-fold [3]. Firstly to understand or model the random (stochastic) mechanism that gives rise to an observed series and secondly to predict or forecast the future values of a series based on the history of that series and, possibly, other related series or factors.

Table 1: Bearing selection-fitting matrix.

Dimension Table

Engine Block: Main Bearing		
Dia = 59 mm	Upp Tol +0.024 µm	Lwr Tol -0 µm
Mark	Range (µm)	
A	0	+6
B	+6	+12
C	+12	+18
D	+18	+24

Crankshaft Main Bearing		
Dia = 55 mm	Upp Tol -0.016 µm	Lwr Tol +0.008 µm
Mark	Range (µm)	
1	+8	+4
2	+4	0
3	0	-4
4	-4	-8
5	-8	-12
6	-12	-16

Main Bearing		
Thickness = 2 mm	Upp Tol -0.06 µm	Lwr Tol +0.012 µm
Colour	Range (µm)	
Blue	+12	+9
Black	+9	+6
Brown	+6	+3
Green	+3	0
Yellow	0	-3
Pink	-3	-6

Fitting Matrix

		Main Bearing			
		A	B	C	D
		Pink	Pink/Yellow	Yellow	Green
Engine Block	1	Pink	Pink/Yellow	Yellow	Green
	2	Pink/Yellow	Yellow	Green	Green/Brown
	3	Yellow	Green	Green/Brown	Brown
	4	Green	Green/Brown	Brown	Black
	5	Green/Brown	Brown	Black	Black/Blue
	6	Brown	Black	Black/Blue	Blue

Time series are generally classified as either stationary or non-stationary. Simplistically, stationary time series process exhibit invariant properties over time with respect to the mean and variance of the series. Conversely, for non-stationary time series the mean, variance or both will change over the trajectory of the time series. Stationary time series have the advantage of representation by analytical models against which forecasts can be produced. Non stationary models through a process of differencing can be reduced to a stationary time series and are so open to analysis applied to stationary processes [4].

An additional method of time series analysis is provided by smoothing the series. Smoothing methods estimate the underlying process signal through creating a smoothed value that is the average of the current plus past observations.

Analysis of Stationary and Non-Stationary Time Series

Stationary time series are classified as having time invariant properties over the trajectory of the time series. It is these time invariant properties that are stationary while the time series itself appears to fluctuating in a random manner. Formally, an observed time series $\{Y(t); t \in T\}$ is weakly stationary if the following properties hold:

1) $E[Y(t)] = \mu, \forall t \in T$

2) $Var[Y(t)] = \sigma^2 < \infty$

3) $Cov(Y_t, Y_{t-j}) = \gamma_j, \forall t, j \in T$

Further, the time series is defined as strictly stationary if in addition to Properties 1-3, if subsequently:

4) The joint distribution of $\{Y(t_1), Y(t_2), ..., Y(t_k)\}$ is identical to $\{T(t_1 + h), Y(t_2 + h), ..., Y(t_k + h)\}$ for any $t_i, h \in T$.

Conversely, a non-stationary time series will fail to meet either or both the conditions of Properties 1 and 2.

Property 3 states that the covariance of lagged values of the time series is dependent on the lag and not time. Subsequently the autocorrelation coefficient ρ_k at lag k is also time invariant and is given by

$$\rho_k = \frac{Cov\left(y_t, y_{t+k}\right)}{Var\left(y_t\right)} = \frac{\gamma_k}{\gamma_0}$$

(1)

The set of all ρ_k, $k = 0,1,2,...,n$ forms the Autocorrelation Function or ACF. The ACF is presented as a graphical plot. Figure 2 provides an example of a stationary time series with the associated ACF diagram. Successive observations of a stationary time series should show little or no correlation and is reflected in the ACF plot showing a rapid decline to zero.

Conversely, the ACF of a non-stationary process will show a gentler decline to zero. Figure 3 replicates the ACF for a non-stationary random walk process.

Stationary time series models are well defined throughout the time series literature where a full treatment of their structure can be found. Representative references include [5] -[8] .

A consistent view of a time series is that of a process consisting of two components a signal and noise. The signal is effectively a pattern generated by the dynamics of the underlying process generating the time series. The noise is anything that perturbs the signal away from its true trajectory [7]. If the noise results in a time series that consists of uncorrelated observations with a constant variance, the time series is defined as a white noise process. Stationary time models are always white noise processes where the noise is represented by a sequence of error or shock terms $\{\varepsilon_t\}$ where $\varepsilon_t \sim N(0,\sigma)$. The time series models applicable to stationary and non-stationary time series are listed in Table 2.

From Table 2, it is seen that by setting the p, d, and q parameters to zero as appropriate, the MA(q), AR(p) and ARMA(p, q) processes can be presented as sub processes of the ARIMA(p, d, q) process and is illustrated in the hierarchy presented in Figure 4. The stationary processes can be thought of as ARIMA processes that do not require differencing.

Table 2: Description of time series models

---

1) Moving Average Process MA (q): The time series y_t is defined as the sum of the process mean and the current shock value plus a weighted sum of the previous "q" past shock values.

2) Autoregressive Process AR (p): The time series y_t is presented as a linear dependence of weighted "p" observed past values summed with the current shock value and a constant.

3) Autoregressive Moving Average ARMA (p, q): The time series y_t is presented as a mixture of both moving average and autoregressive terms. ARMA (p, q) processes require fewer parameters when compared to the AR or MA process (Chatfield, 2006).

4) Autoregressive Integrated Moving Average ARIMA (p, d, q): A non-stationary time series is transformed into a stationary time series through a process of differencing. The ARIMA process differences a time series at most d times to obtain a stationary ARMA (p, q) process.

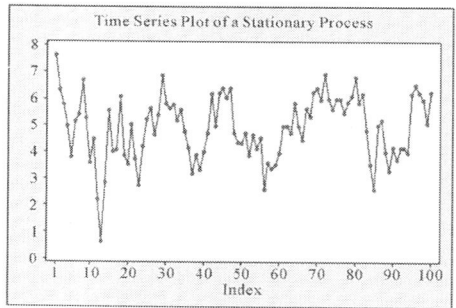 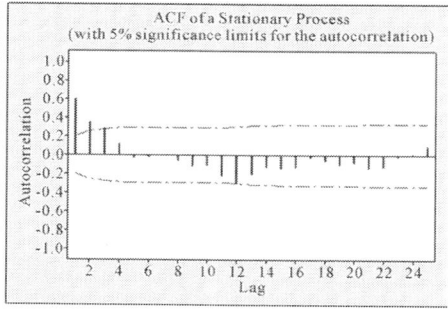

Figure 2: Stationary time series—example ACF.

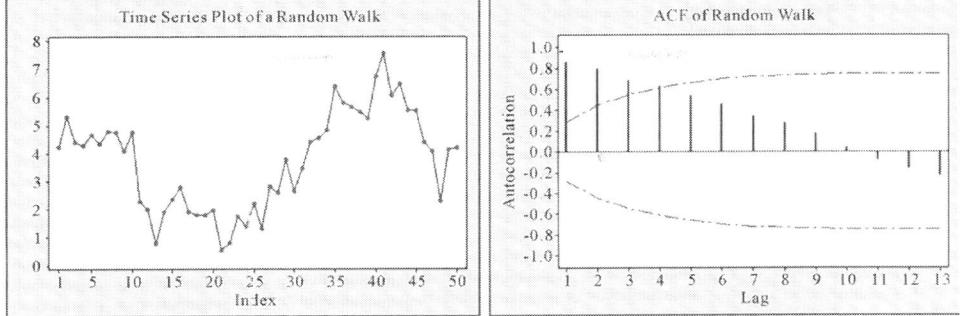

Figure 3: Non-stationary time series example ACF (Random walk).

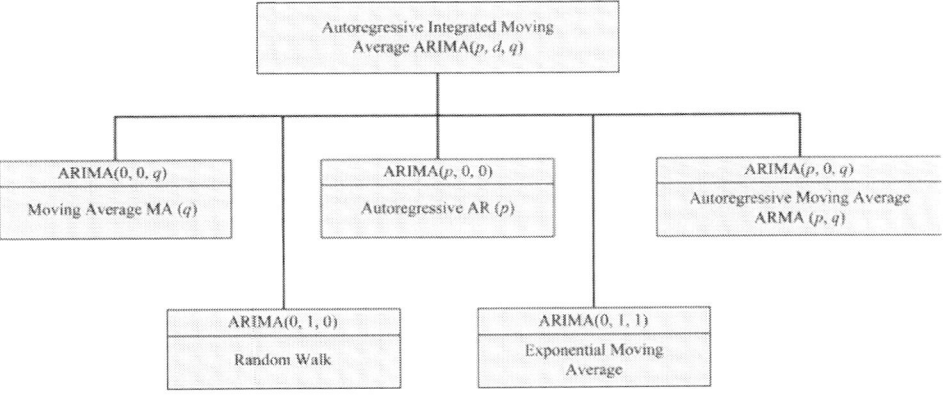

Figure 4: Time series model hierarchy.

Moreover, the ARIMA process reveals time series models that are not open to analysis by the stationary representations. Two such models are included in Figure 4, the random walk model, ARIMA (0, 1, 0) and the exponential moving average model, ARIMA(0, 1, 1).

Table 3 returns the modelling processes that are applied to stationary and non-stationary time series.

Table 3: Synthesis of time series models.

Process	Model	Stationary Condition
MA(q)	$$y_t = \mu + \varepsilon_t - \theta_1\varepsilon_{t-1} - \theta_2\varepsilon_{t-2} - \ldots - \theta_q\varepsilon_{t-q}$$ $$y_t = \mu + \Theta(B)\varepsilon_t$$	None: MA (q) process is always stationary.
AR(p)	$$y_t = \mu + \phi_1 y_{t-1} + \phi_2 y_{t-2} + \ldots + \phi_p y_{t-p} + \varepsilon_t$$ $$\Phi(B)y_t = \delta + \varepsilon_t$$	$$\sum_{i=1}^{p}\phi_t < 1$$
ARMA(p, q)	$$y_t = \mu + \sum_{i=1}^{p}\phi_1 y_{t-i} + \varepsilon_t - \sum_{i=1}^{q}\phi_i\varepsilon_{t-i}$$ $$\Phi(B)y_t = \delta + \Theta(B)\varepsilon_t$$	AR (p) component is stationary.
ARIMA(p, d, q)	$$\Phi(B)(1-B)^d y_t = \delta + \Theta(B)\varepsilon_t$$	The AR (p) component is stationary after the series is differenced d times.

The equations of the models are replicated in their standard form and in terms of the backshift operator B (see Appendix 1). Time series models expressed in terms of the backshift operator are more compact and enable a representation for the ARIMA process that has no standard form equivalent.

Model Identification

Initially, inspection of the ACF of a time series is necessary to determine if the series is stationary or will require differencing. There is no analytical method to determine the order of differencing though empirical evidence suggests that generally first order differencing (d = 1) is sufficient and occasionally second order differencing (d = 2) should be enough to achieve a stationary series [7].

The ACF is also useful to indicate the order of the MA (q) process as it can be shown that for $k > q$ $\rho_k = 0$ Hence the ACF will cut off at lag q for the MA (q) process. The ACF of the AR (p) and ARMA (p, q) processes both exhibit exponential decay and damped sinusoidal patterns. As the form of the ACF's for these processes is indistinguishable, identification of the process is provided by the Partial Autocorrelation Coefficient (PACF).

The properties of the PACF are discussed extensively in the time series literature and in particular a comprehensive review provided in [9] . Descriptively, the PACF quantifies the correlation between two variables that is not explained by their mutual correlations to other variables of interest. In an autoregressive process, each term is dependent on a linear combination of the preceding terms. In evaluating the autocorrelation coefficient ρ_k, the term y_t is correlated to y_{t-k}. However, y_t is dependent on y_{t-1} which in turn is dependent on y_{t-2} and the dependency propagates throughout the time series to y_{t-k}. Consequently, the correlation at the initial lag of the time series propagates throughout the series. The PACF evaluates the correlation between x_t is and x_{t-k} through removing the propagation effect.

The PACF can be calculated for any stationary time series. For an AR (p) process, it can be shown that the PACF cuts off at lag p. The PACF for both the MA (q) and ARMA (p, q) process the PACF is a combination of damped sinusoidal and exponential decay.

The structure of a stationary time series is determined through inspection of both the ACF and PACF diagrams of the series. Table 4 presents the classification of the stationary time behaviour (adapted from [7] [8]).

Neither the ACF nor PACF yield any useful information with respect to identifying the order of the ARMA (p, q) process. Though there are additional methods that can aid the identification of the required order of the process [7] [8] including the extended sample autocorrelation function (ESACF), the generalised sample partial autocorrelation function (GPACF), and inverse autocorrelation function (IACF) as suitable methods for determining the order of the ARMA model. However, with respect to investigating the structure of a time series both sets of authors agree that with the availability of statistical software packages it is easier to investigate a range of models with various orders to identify the appropriate model and forego the need to apply these additional methods.

The estimation of the parameter coefficients ϕ_i and θ_i can be estimated through Maximum Likelihood Estimation (MLE). An extensive explanation of the MLE application to estimate the parameters of each the models

Table 4: Classification of stationary time series behavior.

Model	ACF	PACF
MA(q)	Cuts off after lag q	Infinite exponential decay and/or damped sinusoid— tailing off
AR(p)	Infinite exponential decay and/or damped sinusoid— tailing off	Finite-cuts off after lag p
ARMA(p,q)	Infinite exponential decay and/or damped sinusoid— tailing off	Infinite exponential decay and/or damped sinusoid— tailing off

Listed in Table 2 is provided in [10]. Additionally the parameters of the AR (p) process can be calculated with the Yule Walker Equations [4] [9] .

The robustness of a derived model is assessed through residual analysis of the model. For the ARMA (p, q) process the residual are obtained from

$$\hat{\varepsilon}_t = y_t - \left(\hat{\delta} \sum_{i=1}^{p} \hat{\phi}_i y_{t-1} - \sum_{i=1}^{q} \hat{\theta}_i \varepsilon_{t-1} \right)$$

(2)

If the residual values exhibit white noise behaviour, the ACF and PACF diagrams of the residual values should not show any discernable pattern then the appropriate model has been chosen and the correct orders of p and q have been correctly identified.

Generation of Forecasts

Establishing a model that describes the structure of an observed time series enables meaningful forecasts to be drawn from the model. Forecasting methods for each of the models in Table 2 are succinctly described in [11] and are defined as follows:

AR (p) Process:

The forecast for the AR (p) process is based on the property that the expectation of the error terms ε_i are zero. The forecast is developed iteratively from the previous observation to create a one step ahead forecast. The step ahead is denoted by

$$\tau = 1 : \hat{y}_t(1) = \delta + \phi_1 y_{t-i} + \phi_2 y_{t-i} + \dots + \phi_p y_{t-i+1}$$

$$\tau = 2 : \hat{y}_t(2) = \delta + \phi_1 \hat{y}_t(1) + \phi_2 y_{t-2} + \dots + \phi_p y_{t-p+2}$$

At each successive iteration, the most recent observation drops out of the forecast and replaced by the previous forecast value. At $\tau > p$, each term is a forecasted value and continuing the iteration process, the forecast converges onto the mean of the AR(p) process.

MA (q) Process:

The expectation of the ε_i terms in the MA(q) process follow a white noise process and so the expectation of all future values of ε_i, $i > 0$ is zero. Hence for $t > q$ the forecast of a MA (q) process is just the mean value of the process, μ.

ARMA (p, q) Process:

The forecast for the ARMA (p, q) process is the combination of the results from the pure AR (p) and MA (q) processes. For $t > q$, the error terms completely drop out of the forecast.

Smoothing Methods

Smoothing methods provide a class of time series models that have the effect of reducing the random variation of a time series with the effect of exposing the process signal within the time series. A variety of smoothing models that average the series continually over a moving span of observed values or fit a polynomial approximation to a restricted interval of the series are presented in [12].

Of the smoothing techniques available, the exponential smoothing model has proved useful due to its simplicity of application. First order exponential smoothing is defined by the following recurrence relation:

$$\tilde{y}_t = \lambda y_t + (1 - \lambda)\tilde{y}_{t-1}$$

$$(3)$$

Where λ is the smoothing factor $(0 < \lambda < 1)$?

Effectively, first order exponential smoothing is a linear combination of the current observation plus the discounted sum of all previous observations due to the smoothing factor λ. Moving back through the recursive relation $(1 - \lambda)$ geometrically decays and so the older observations have a diminishing contribution to the smoothed estimation of the current value.

Though the recursive relation defined in Equation (3) can be expressed as an ARIMA (0, 1, 1) model, the method was initially developed from first principles in [13] as a means of forecasting inventory.

First order exponential smoothing will for trending data lag behind an increasing trend and lead a decreasing trend. Second order exponential smoothing overcomes this problem as does increasing the value of the smoothing factor [7]. However, providing the time series is showing no systematic trend, first order exponential smoothing is an adequate model to analyse a time series [6].

The forecast generated from first order exponential smoothing is just the value of the smoothed current value and in principle would extend over all future values. However, as more observed values become available it makes sense to update the forecast.

SIMULATION APPLICATION TO EVALUATE CONSUMPTION RATE (MONTE CARLO METHOD)

Frequently in real world scenarios due to the complexity of the system under investigation it may not be possible to evaluate the systems behaviour by applying analytical methods. Under such conditions an alternative approach to model such system is through creating a simulation. Succinctly, simulation methods provide an alternative approach to studying system behaviour through creating an artificial replication or imitation of the real world system [14] . The applications where simulation methods may be useful is extensive and include diverse disciplines such as manufacturing systems, flight simulation, construction, healthcare, military and economics [15] [16] .

Systems or processes that can be modelled through an underlying probability distribution are open to simulation through the Monte Carlo method [17]. The method simulates the behaviour of a system by taking repeated sets of random numbers from the underlying probability distribution of the process under investigation. The development of

Table 5: Monte Carlo method process stages (Adapted from [17]).

Stage 1 Define a distribution of possible inputs for each input random variable: Requires recognition of the underlying probability distribution of the process. This may be directly apparent or may require empirical observation of the process under investigation.

Stage 2 Generate outputs randomly from those distributions: Requires the selection of an appropriate random number generator to model the observed probability distribution. Random number generators are generally available in most statistical software packages and Micro Soft Excel.

Stage 3 Perform a deterministic computation using that set of outputs: Involves computing the desired output variable or variables from the generated random numbers.

Stage 4 Aggregate the results of the individual computations into the final result: The aggregation process is dependent on the specific simulation but could be as straightforward as computing the average of the simulated results.

---

the method is attributed to the work of Stanislaw Ulam and John von Neumann during the late 1940's to aid their work in atomic physics for the development of nuclear weapons. Analytical methods were proving difficult if not impossible to apply and Ulam turned to random experimentation to elicit system attributes and behaviour [18].

A specific application of the Monte Carlo method is dependent on the nature of the underlying probability distribution of the system or process under investigation. However the method application is consistent and will follow the steps outlined in Table 5.

Application of Monte Carlo Method to Shell Consumption Case Study

Visual inspection of the time series trajectories of the shell consumption does not yield an obvious probability distribution so Stage 1 of the Monte Carlo process begins by examining the histograms of the observed series.

Figure 5 returns the histograms of the consumption of the Green and Yellow Shell consumption for the full trajectory of observed values. The histograms for each shell indicate the data is negatively skewed and so drawing random numbers from a skewed normal distribution for the simulation is deemed appropriate. It can also be shown that the distributions for shorter time spans of the observed values are also skewed.

Stages 2 to 4 of the Monte Carlo process are embedded in a Visual Basic for Applications (VBA) program created in Microsoft Excel to carry out the simulation. Figure 6 provides a schematic of the VBA simulation programme.

The VBA programme incorporates a skewed normal random number generation algorithm developed in [19] that generates random numbers based on the skew value of sample input data. The Excel spread sheet contains the consumption history for each shell. Upon invoking the simulation, the required sample and forecast periods

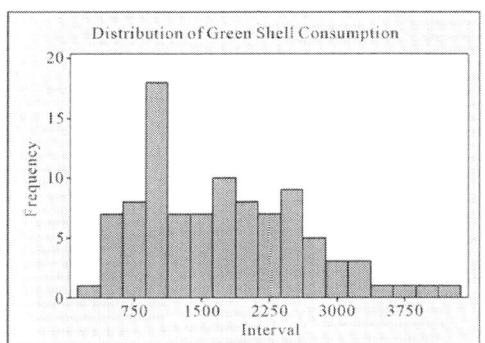 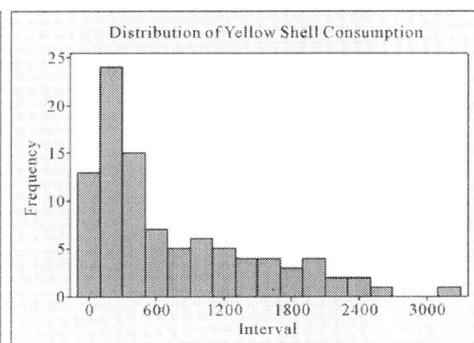

Figure 5: Histograms of green and yellow shell consumption.

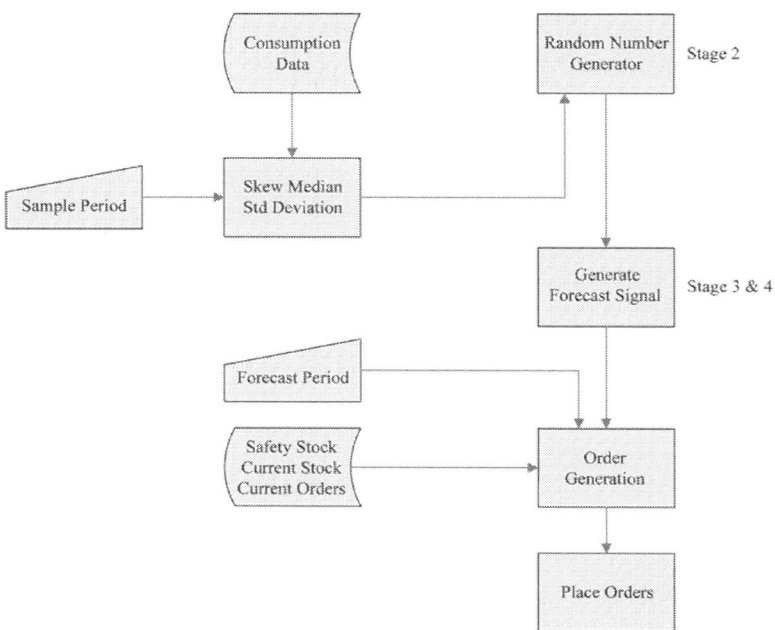

Figure 6: Schematic of VBA simulation programme.

Are read from input data in the spread sheet. For each shell, against the sample period, the skew, median and standard deviation values are calculated and input into the random number generator. For each week of the sample period, the generator calculates 100 positive random numbers to create an average simulated consumption for each week. Since it is impossible to have negative consumption, negative random numbers are rejected. Against each weekly average, an overall average is computed to create the forecast signal.

The random number generator satisfies Stage 2 of Monte Carlo process, while Stages 3 and 4 are satisfied through computation of the forecast signal.

Additional functionality is provided in the simulation tool that over the forecast period will generate orders taking into account current stock levels, safety levels and orders already placed.

GENERATION OF FORECASTS FOR THE BEARING SHELL CONSUMPTION

The time series and the Monte Carlo methods described in the preceding sections are applied to the historical shell bearing consumption usage to generate forecasts to create orders to satisfy future engine build requirements. The forecasts of each model are compared to determine if there is either a consistent or significant differences between each method.

ARIMA (p, d, q)

The purpose of the ARIMA (p, d, q) process is to establish the underlying model that describes the time series through specifying the p and q parameters once the appropriate order of differencing d has identified a stationary process. Model identification is primarily based on inspection of the ACF and PACF diagrams and referencing the Classification Table (Table 4). The robustness of the forecast generated from the identified model is determined by Residual Analysis where primarily if the ACF and PACF diagrams of the residual values show no discernible pattern, the forecast can be assessed as robust [2] -[11] . It is considered that the identification of a time series model requires both judgement and experience and it is recommended that the iterative model building process illustrated in Figure 7 is followed [8] [9] .

The forecasting process is illustrated against the Green Shell consumption replicated in Figure 1. The analysis is carried out using the Minitab® statistical package. The ACF and PACF diagrams are replicated in Figure 8 and with reference to Table 4 indicates that the time series follows an AR(1) process.

Table 6 returns the output generated against 40 observations. The ACF and PACF diagrams of the residual values are returned in Figure 9 and do not show any discernible pattern. The Chi Square statistics appear to be

Table 6: Model estimation for green usage.

ARIMA Model: Green				
Final Estimates of Parameters				
Type	Coef	SE Coef	T	P
AR 1	0.6940	0.1168	5.94	0.000
Constant	368.1	104.8	3.51	0.001
Mean	1203.0	342.5		
Number of observations: 40				

Residuals: SS = 16,691,055 (backforecasts excluded)

MS = 439,238; DF = 38

Modified Box-Pierce (Ljung-Box) Chi-Square statistic				
Lag	12	24	36	8
Chi-Square	14.9	21.1	23.9	*
DF	10	22	34	*
P-Value	0.135	0.518	0.902	*

Low and diminish as the number of lags increases.

Evidence therefore exists to support that the residuals follow a white noise process and the AR (1) is a robust representation of the observed time series.

The AR (1) process is defined by

$$y_t = 368.1 + 0.694y_{t-1}$$

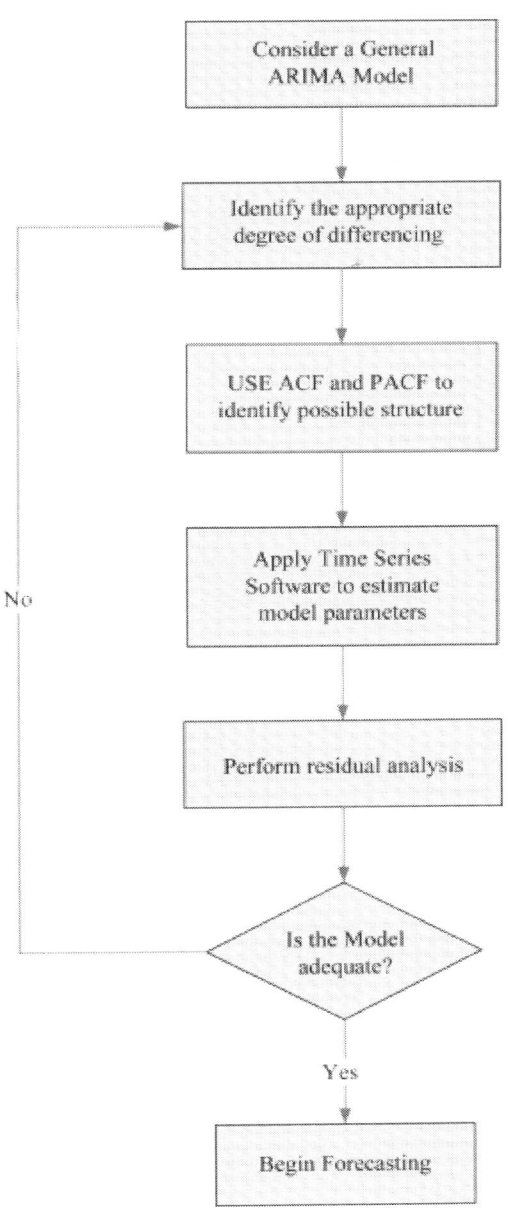

Figure 7: ARIMA model building stages (Adapted from [8] [9]).

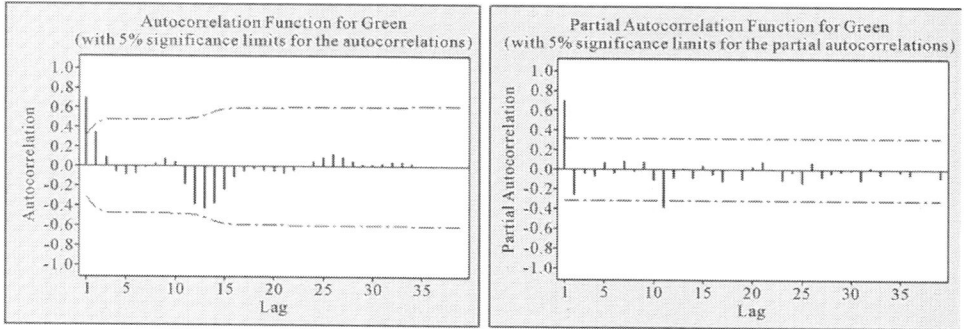

Figure 8: ACF and PACF of green shell usage.

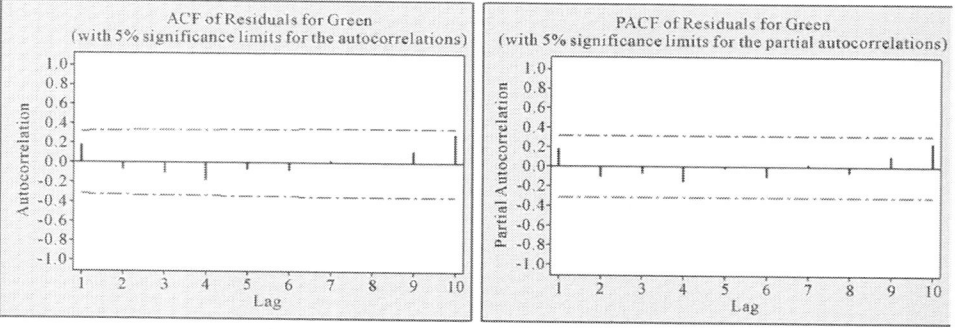

Figure 9: ACF and PACF for residuals of the AR (1) process.

The forecast for the process will converge onto the process mean $\mu = 1203$.

The procedure applied to the Yellow shell consumption also resulted in an AR (1) process. Though the details are omitted, the AR (1) process is defined by

$$y_t = 86.73 + 0.8665y_{t-1}$$

The forecast for the process will converge onto the process mean $\mu = 649$.

Exponential Smoothing

The application the first order exponential smoothing method requires a choice of smoothing factor λ and the number of observations to smooth against. There is no analytical method to determine the optimum choice of smoothing factor and it is necessary to investigate various levels of λ and choose the smoothing factor that minimises the squared sum of the forecast errors e_t defined by

$$SS_E(\lambda) = \sum_{t=1}^{T} e_t^2 .$$

Similarly there is no method to determine the optimum number of observations to forecast against. The influence of past the observations decays geometrically over time and the impact of the decay can be evaluated against a set of observations.

Table 7 returns gererated forecasts for the varying levels of λ against both the Green and Yellow shell data where the full set of 109 observations are used. Due to the magnitude of errors, the square root of $SS_E(\lambda)$ is returned in Table 7.

There is a distinct contrast between the two forecasts. Forecast accuracy for the Green consumption increases as λ increases and is a consequence of the volatility of the series. For the Yellow consumption forecast accuracy is assured at $\lambda = 0.4$, where $SS_E(\lambda)$ takes the minimum value. Evident for both time series is that the forecast is influenced by the level of the smoothing factor λ. The choice of smoothing factor for the Yellow forecast is defined by the existence of a minimum $SS_E(\lambda)$. The choice of smoothing factor for the Green forecast is not so clear. By inspection of Table 6, choosing $\lambda = 0.6$ would appear appropriate as the forecast reduces at $\lambda = 0.8$ drops slightly.

Table 8 returns the forecast generated at different observation levels for specific choices of smoothing factor λ.

Variation in the forecast level only becomes apparent at the lower number of observations and the variation is insignificant relative to the level of each forecast and the width of the confidence intervals.

The evidence from Table 7 and Table 8 suggests that the level of smoothing factor is critical with respect to generating a robust forecast while the choice of the number of observations to forecast against is not critical.

Monte Carlo Simulation

Table 9 returns output from the simulation programme for a range of sample periods for both the Green and Yellow shell consumption. What is clear from both sets of simulations is that for any sample period the simulated forecast signals are consistent. For the Green simulation as the sample period increases, volatility of the consumption is exposed to the simulation. As the sample period increases, the standard deviation increases resulting in an inflated forecast relative to the sample mean. For the Yellow simulation, as the sample period is Extended, the standard deviation has not increased significantly reflecting that over extended sample periods the distribution of observations has a greater consistency.

The simulation process is sensitive to excessive volatility resulting in an inflated forecast signal. In principle the effect of the volatility can be overcome by reducing an excessive observation to say within one

Table 7: Forecasts generated against smoothing factor level.

	Green Forecast Summary (N = 109)						Yellow Forecast Summary (N = 109)				
λ	Forecast	LCL	UCL	Con Int	Sqrt $(SS_E(λ))$	λ	Forecast	LCL	UCL	Con Int	Sqrt $(SS_E(λ))$
0.2	1095.82	−607	2798	3405	9352	0.2	502.18	−246	1251	1497	4273
0.4	1209.07	−218	2636	2855	8048	0.4	644.26	−13	1302	1315	3958
0.6	1265.79	−16	2547	2563	7423	0.6	752.99	107	1399	1292	4034
0.8	1262.91	37	2489	2452	7112	0.8	824.60	166	1483	1318	4256

Table 8: Forecasts generated against number of observations.

Green					Yellow			
$\lambda = 0.6$	Forecast	LCL	UCL	Con Int	$\lambda = 0.4$	Forecast	LCL	UCL	Con Int
N = 109	1265.789	−15.6988	2547.277	2562.975	N = 109	644.2553	−13.0679	1301.579	1314.646
N = 50	1265.789	−98.4743	2630.052	2728.526	N = 50	644.2553	100.0596	1188.451	1088.391
N = 20	1265.789	251.1019	2280.476	2029.374	N = 20	644.2454	171.9441	1116.547	944.6025
N = 10	1265.889	717.1508	1814.628	1097.477	N = 10	643.8595	21.71238	1266.007	1244.294

Table 9: Simulated forecast output.

Green Simulation

Period	Mean	St Dev	Simulated Forecast Signal				
N = 10	1055.70	271.31	1014	1016	1010	1019	1021
N = 15	933.53	483.71	921	918	924	917	926
N = 20	1013.30	543.61	1271	1271	1269	1277	1264
N = 25	1289.44	907.56	1760	1773	1759	1769	1779

Yellow Simulation

Period	Mean	St Dev	Simulated Forecast Signal				
N = 10	461.00	279.72	475	478	486	475	481
N = 15	344.40	292.29	442	445	448	454	445
N = 20	328.95	257.37	422	419	415	416	425
N = 25	391.08	303.48	503	498	501	507	496

or two standard deviations of the sample mean. The difficulty in this approach is creating a consistent rule that would apply to all excessive observations. Setting a trigger point at for example 2.5 standard deviations above the sample, an observation at 2.6 standard deviations would be excessively reduced while an observation at 3.5 standard deviations may not be reduced enough.

Comparison of Forecasting Methods

Table 10 returns the observations of the shell consumption for the following 12 weeks beyond the original observations and the respective graphs of the time series is returned in Figure 10.

A summary of the output from each method is presented in Table 11 where the totals of each forecast are compared to the total of the observed values over the forecast period.

The forecast signals are reasonably close to one another. Though the simulated forecast signal for the Yellow shell is significantly lower than the signals generated by the time series methods, the total forecast is in excess of the observed consumption over the forecast period. The consistency of the forecast signals imply that each modelling processes is robust so indicating the reliability of the signals to generate purchase orders for the shells.

Table 10: Observed shell consumption over forecast period.

	1	2	3	4	5	6	7	8	9	10	11	12	Total
Green	1433	1449	234	305	319	308	362	721	795	877	791	941	8535
Yellow	1115	509	111	491	445	374	387	603	457	432	89	59	5072

Table 11: Summary of forecast output.

Method	Shell	Forecast Signal	Total Forecast	Observed Consumption	Excess
Autoregressive	Green	1203	14,436	8535	5901
	Yellow	649	7788	5072	2716
Exponential Smoothing	Green	1265	15,180	8535	6645
	Yellow	644	7728	5072	2656
Simulation	Green	1275	15,300	8535	6765
	Yellow	475	5700	5072	628

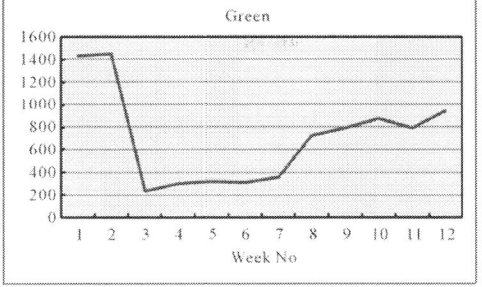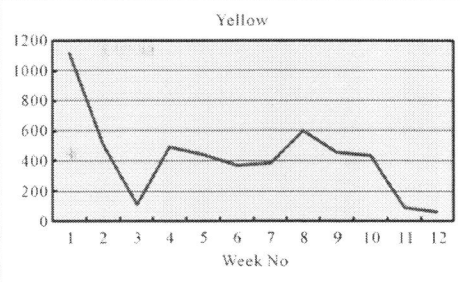

Figure 10: Time series observations over forecast period.

CONCLUSIONS

Historically, forecasting future demand of the crankshaft shells has proven exceptionally difficult. The purchasing professionals within the case study environment having no experience of applying formal forecasting methods placed orders on the crankshaft shell suppliers that were effectively a "best guess". To mitigate the need to guess, formal time series analysis was assessed as a suitable approach to modelling demand along with a Monte Carlo simulation method coded in Microsoft Excel VBA.

The forecasts generated by the exponential smoothing and autoregressive process are remarkably close. The simulation process though proving to be sensitive to volatile consumption with judgement of the physical time series, an appropriate forecast signal can be found.

The advantage of the exponential smoothing method lies in its simplicity of execution. The method is open to coding in VBA and would therefore negate the need for specialised software. Though for a volatile time series, it may not be possible to find a smoothing factor that will maximise forecast accuracy. However, choosing a smoothing factor between 0.5 and 0.7 should provide an appropriate forecast signal.

The ARIMA process requires interaction between the user and the modelling process as the ACF and PACF have to be inspected to determine the structure of the model. Though in this study, the autoregressive models were basic AR (1) models, future trajectories may follow more complex ARIMA models. It is recommended that at least 30 observations of a time series are needed to enable the generation of a meaningful forecast. In the current study, there is over two years of data available so the restriction did not apply. However, for the build of a new engine, the application of the ARIMA model would be impeded until there are enough observations to forecast against. Empirical evidence from this study confirms that the exponential smoothing model can produce meaningful forecasts with at least 10 observations and would therefore provide a more robust method of forecasting with limited data sets.

The simulation method provides a seamless process to generate forecast signals and offers additional functionality to generate orders. Execution of the simulation for the complete shell requirement is efficient and completes generally in less than two minutes. The simulation is sensitive to volatile demand and can over inflate the forecast signal.

Monte Carlo methods have proved effective to model a diverse range of complex applications. While the method is consistent, the execution of the method to a specific application has to be tailored to that application. The functionality of the simulation method created within this study is restricted to modelling the forecast consumption of the crankshaft shells. Conversely, the time series methods are universally applicable.

Inspection of Table 11 confirms that for each of methods, the forecast signals are close to each other and so with respect to forecast accuracy there is no one best method. Moreover, since the forecast signals are close to one another, the purchasing professionals are confident that a robust forecast signal is being generated.

Currently, the appropriate choice of forecasting tool is the simulation method as it provides a seamless process not only to generate the forecast signal but also generate the orders. Building a VBA programme around the exponential smoothing process should in principle provide the functionality provided by the simulation method.

Inventory profiles exhibited by the consumption of the crankshaft shells are rare within the discipline of Operations Management. Conventional though rigorous methods of inventory control that include Materials Requirement Planning and Just-in-Time Kanban systems do not apply to the procurement of the crankshaft shells. Moreover, due to the rarity of such inventory profiles, there does not appear to be any significant research into this unique area of inventory management. The application of the Monte Carlo simulation method and time series analysis begins to close this gap.

Appendix 1: Backshift and Differencing Operators

Frequently within the time series literature, time series are presented using difference operator denoted by ∇ and the lag or backshift operator B. Kirchgässner et al. (2013) succinctly define the properties of both operators. The essential properties of the operators are replicated as follows:

First order differencing is expressed using the difference operator ∇ such that

$\nabla y_t = y_t - y_{t-1}$ Second order differencing is expressed as

$$\nabla^2 y_t = \nabla(y_t - y_{t-1}) = y_t - 2y_{t-1} + y_{t-2}$$

The backshift operator B has the effect of "delaying" a time series by one period, such that

$$By_t = y_{t-1}$$

Applying the backshift operator to y_{t-1}, the following is obtained

$$By_{t-1} = B(By_t) = B^2 y_t$$

In general

$$B^k y_t = y_{t-k} \tag{A1}$$

Applying property (A1) to a time series of the form

$$y_t = \varepsilon_t + \theta_1 \varepsilon_{t-1} + \theta_2 \varepsilon_{t-2} + \ldots + \theta_q \varepsilon_{t-q}$$

Yields a polynomial in B such that

$$y_t = \left(1 + \theta_1 B + \theta_2 B^2 + \ldots + \theta_q B^q\right)\varepsilon_t = \Theta(B)\varepsilon_t$$

The backshift operator can be related to a first order difference in the following way

$$\nabla y_t = y_t - y_{t-1} = y_t - By_t = (1 - B)y_t$$

Second order differencing

$$\nabla^2 y_t = \nabla(\nabla y_t) = (1 - B)(1 - B)y_t = (1 - B)^2 y_t$$

If a series is differenced d times then it can be shown that

$$\nabla^2 y_t = (1 - B)^d y_t$$

REFERENCES

1. Hillier, V.A.W. and Coombes, P. (2004) Fundamentals of Motor Vehicle Technology Book 1. 5th Edition, Nelson Thornes Ltd, Cheltenham.
2. Woodward, W.A., Gray, H.L. and Elliot, A.C. (2012) Applied Time Series Analysis. CRC Press, Boca Baton.
3. Cryer, J.D. and Chan, K.S. (2008) Time Series Analysis: With Applications in R. Springer. Science + Business Media, London.
4. Brockwell, P.J. and Davis, R.A. (2010) Introduction to Time Series and Forecasting. 2nd Edition, Springer Texts in Statistics, New York.

5. Peña, D., Tiao, G.C. and Tsay, R.S. (2000) A Course in Time Series Analysis. Wiley-Interscience, New York.
6. Chatfield, C. (2006) the Analysis of Time Series, an Introduction. 6th Edition, Chapman and Hall, London.
7. Montgomery, D.C., Jennings, C.L. and Kulachi, M. (2008) Introduction to Time Series Analysis and Forecasting. WileyBlackwell, New Jersey.
8. Bisgard, S. and Kulachi, M. (2011) Time Series Analysis and Forecasting by Example. Wiley & Sons Inc., New York. http://dx.doi.org/10.1002/9781118056943
9. Box, G.E.P., Jenkins, G.M. and Reinsel, G.C. (2008) Time Series Analysis. Forecasting and Control. 4th Edition, Wiley & Sons Inc., New Jersey.
10. Hamilton, J.D. (1994) Time Series Analysis. Princeton University Press, Princeton.
11. Kirchgässner, D., Wolters, J. and Hassler, U. (2013) Introduction to Modern Time Series Analysis. 2nd Edition, Springer Heidelberg, New York. http://dx.doi.org/10.1007/978-3-642-33436-8
12. Kendall, M. (1976) Time Series. 2nd Edition, Charles Griffin and Co Ltd., London and High Wycombe.
13. Brown, R.G. (1956) Exponential Smoothing for Predicting Demand.http://legacy.library.ucsf.edu/tid/dae94e00
14. Robinson, S. (2004) Simulation: The Practice of Model Development and Use. John Wiley and Sons, Chichester.
15. Banks, J., Carson, J.S. and Nelson, B.L. (1996) Discrete-Event System Simulation. 2nd Edition, Prentice-Hall, Upper Saddle River.
16. Singh, V.P. (2009) System Modelling and Simulation. New Age International Publishers, New Delhi.
17. Sokolowski, J.A. (2010) Monte Carlo Simulation. In: Sokolowski, J.A. and Banks, C.M., Eds., Modelling and Simulation Fundamentals: Theoretical Underpinnings and Practical Domains, Wiley & Sons Inc., New Jersey, 131-145.http://dx.doi.org/10.1002/9780470590621.ch5
18. Metropolis, N. and Ulam, S. (1949) the Monte Carlo Method. Journal of the American Statistical Association, 44, 335-341. http://www.amstat.org/publications/journals.cfmhttp://dx.doi.org/10.1080/01621459.1949.10483310
19. Azzalini, A. (2008) the Skew-Normal Probability Distribution.http://azzalini.stat.unipd.it/SN/

CITATION

Davies, R., Coole, T. and Osipyw, D. (2014) The Application of Time Series Modelling and Monte Carlo Simulation: Forecasting Volatile Inventory Requirements. Applied Mathematics, 5, 1152-1168. doi: 10.4236/am.2014.58108.

Hurst's Memory For Chaotic, Tree Ring, And SOI Series

Byung-Sik Kim[1], Hung-Soo Kim[2], and Sun-Hong Min[1*]

[1]Department of Urban Environmental Prevention Engineering, School of Disaster Prevention, Kangwon National University, Gangwon, South Korea
[2]Department of Civil Engineering, Inha University, Incheon, South Korea

ABSTRACT

Hurst's memory that roots in early work of the British hydrologist H.E. Hurst remains an open problem in stochastic hydrology. Today, the Hurst analysis is widely used for the hydrological studies for the memory and characteristics of time series and many methodologies have been developed for the analysis. So, there are many different techniques for the estimation of the Hurst exponent (H). However, the techniques can produce different characteristics for the persistence of a time series each other. This study uses several techniques such as adjusted range, rescaled range (RR) analysis, modified rescaled range (MRR) analysis, 1/f power spectral density analysis, Maximum Likelihood Estimation (MLE), detrended fluctuations analysis (DFA), and aggregated variance time (AVT) method for the Hurst exponent estimation. The generated time series from chaos and stochastic systems are analyzed for the comparative study of the techniques. Then, this study discusses the advantages and disadvantages of the techniques and also the limitations of them. We found that DFA is the most appropriate technique for the Hurst exponent estimation for both the short term memory and long term memory. We analyze the SOI (Southern Oscillations Index) and 6 tree-ring series for USA sites by means of DFA and the BDS statistic is used for nonlinearity test of the series. From the results, we found that SOI series is nonlinear time series which has a long term memory of H = 0.92. Contrary to earlier work, all the tree ring series are not random from our analysis. A certain tree ring series show a long term memory of H = 0.97 and nonlinear property. Therefore, we can say that the SOI series has the properties of long memory and nonlinearity and tree ring series could also show long memory and non-linearity.

INTRODUCTION

Hydrologic or geophysical time series may have a certain dependent structure in itself. For example, if this month has a monthly stream-flow with high level at a station, next month could also have a high streamflow. This describes that the consecutive values of hydrologic or geophysical time series show self-dependence between the values which has been known as short-range (or short-term) dependence, per-sistence, or memory [1]. analyzed the sequence of annual discharges from the Nile River in Egypt to estimate storage volume. The terms "Hurst phenomenon" and "Joseph effect" (due to Mandelbrot from the biblical story of the "seven years of great abundance" and the "seven years of famine") have been used as alternative names for the [2].

[3] suggested a method called the R/S analysis for detecting long mem-ory and the method allowed the calculation of Hurst's exponent or self-similarity parameter. [3] divided the range by the standard devia-tion of the sample S(n) and called it the "rescaled range" R/S statistic.

Short-term memory implies that the effect of an observation in a time series on the future observations becomes negligible after a short pe-riod of time. Contrary to short-term memory and long-term memory implies that the effect of an observation on future observations remains significant for a long period of time. Following Hurst's approach, the detection of long memory can be done heuristically by estimating its intensity, namely the value of the parameter H, which varies between 0 and 1. The H value equal to 0.5 means absence of long memory.

Hurst exponent is being used in many fields such as physiology, elec-tronics and computer science and used to determine the persistence of signals such as brain wave, an electrocardiogram and electronic wave.

The exponent is also used to investigate the chaotic behavior or self-similarity of the system [4,5]. There are many techniques for the estimation of Hurst exponent such as 1/f power spectral density (PSD) [6], aggregated variance time (AVT) method [7-9], detrended fluctuations analysis (DFA) [10,11], maximum likelihood estimation (MLE) [12].

In this study, we investigate the estimation techniques of Hurst exponent by estimating the exponent for chaotic and stochastic time series and comparing the results for examining the advantages, disadvantages, and limitations of the techniques. We also select an appropriate technique by the examination and apply it to the SOI (southern Oscillation Index) and 6 tree-ring series.

COMPARISON OF HURST EXPONENT ESTIMATION METHODS

Various methods exist for the Hurst exponent estimation and this section discusses the methods by analyzing the data sets generated from the stochastic and chaotic systems. We also discuss the advantages and disadvantages of the methods and the limitations of them through the estimation of Hurst exponents for the data sets from the systems.

Data Used

This study tests the time series with the known Hurst exponents as shown in 1) to 4) in below and the series generated from chaotic system and a nonlinear stochastic model as in 5) to 7). Each series with the size of 1000 is generated from the systems and the time series plots are shown in Figure 1.

1) Gaussian white noise; iid (identified independant distribution); H ~ 0.5

$$x_t = \varepsilon_t \sim N(0,1) \tag{1}$$

where,N: normal distribution with mean 0 and standard deviation 1 t: time 2) Fractional Gaussian noise with long range correlation (FGN); H ~ 0.8

$$x_i = B_H(i+1) - B_H(i) \quad i \geq 1 \tag{2}$$

where,B_H: Fractional Brownian motion.

3) Autoregressive model (AR(1)) with $\rho_1 = 0.7$

$$x_t = \rho_1 x_{t-1} + \varepsilon_t \sim N(0,1) \tag{3}$$

where,P_1: lag-1 autocorrelation coefficient.

ε_t: White noise 4) Fractionally Differenced ARMA (FARIMA); H ~ 0.8; ARIMA (1,0.3.1)

$$\Phi(B)(1-B)^d x_t = \Theta(B)\varepsilon_t \sim N(0,1) \tag{4}$$

where, $\phi(B)$: parameter of AR model

$\Theta(B)$: parameter of MA model

B: back shift operator

d: H–1/2 5) Three torus—quasi-periodic function

$$x_t = \sin\left[\frac{3t}{500}\right] + \sin\left[\frac{3\sqrt{2}t}{250}\right] + \sin\left[\frac{9\sqrt{3}t}{500}\right], \quad t \geq 1 \tag{5}$$

where, t; time

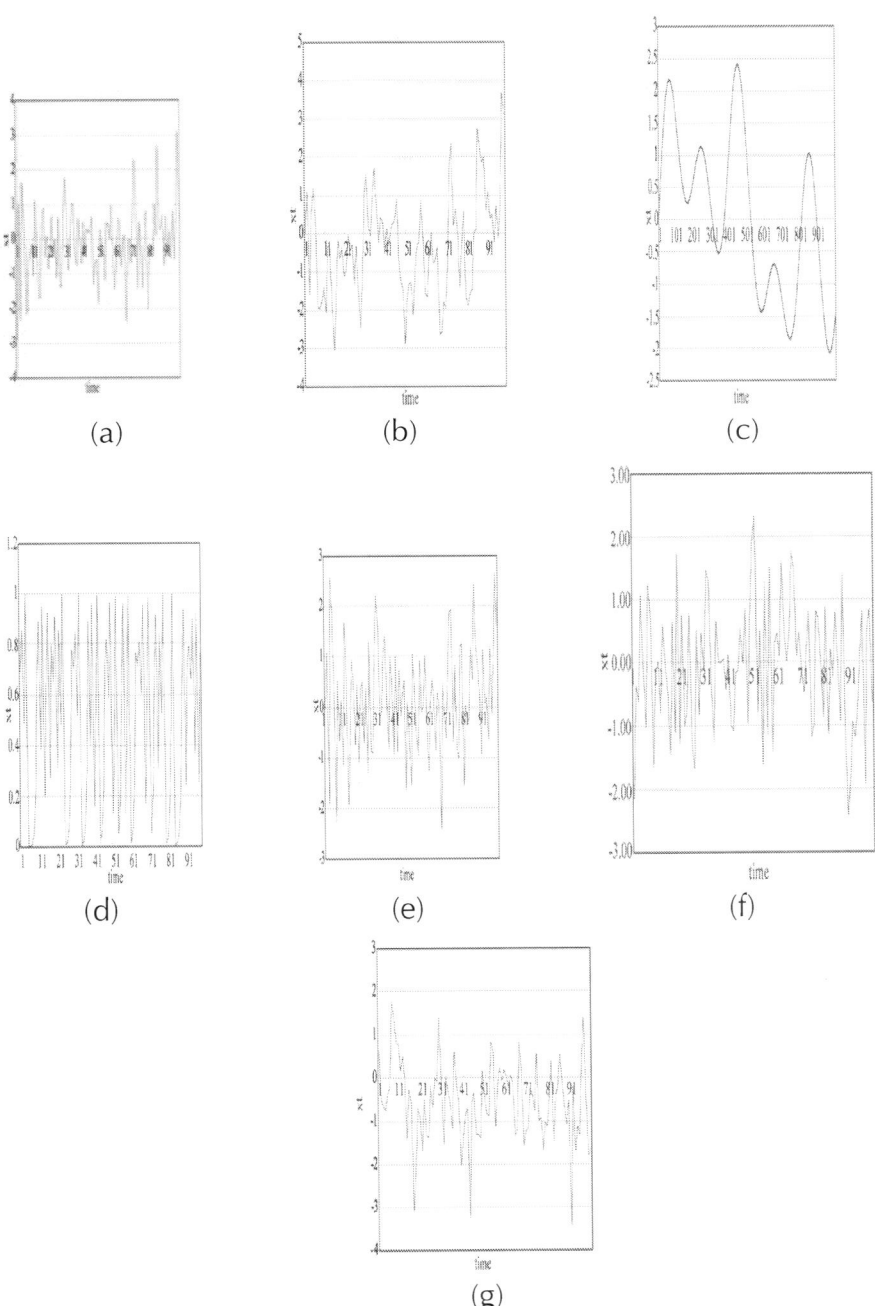

Figure 1: The generated time series from chaotic and stochastic systems. (a) white noise; (b) AR(1); (c) Three torus; (d) Logistic map; (e) TAR(2,1); (f) FARIMA; (g) FGN.

6) Logistic map

$$x_{t-1} = 4x_t\left(1 - x_t\right)$$

(6)

where, t: time 7) Threshold autoregressive model (TAR(2,1))

$$x_t = -0.5x_{t-1} + \varepsilon_t \quad \text{if } x_{t-1} < 1$$
$$x_t = +0.4x_{t-1} + \varepsilon_t \quad \text{if } x_{t-1} \geq 1$$

(7)

where, t: time.

Methods for the Estimation of Hurst Exponent@Nolisttemp#

Adjusted Range

[1], examined several sequences of streamflow and observed that for a sequence of n, 1 ≤ n ≤ N, the adjusted range statistic, denoted by R, is defined as

$$R = \left[\max \sum_{t=1}^{N}\left(X_t - \overline{X_n}\right) - \min \sum_{t=1}^{N}\left(X_t - \overline{X_n}\right) \right]$$

(8)

$$R/\sigma = kN^{\beta} \rightarrow R/\sigma = \left(\frac{N}{2}\right)^{H}$$

(9)

$$H = \frac{\text{Log}\left(R_N\right)}{\text{Log}\left(N/2\right)}$$

(10)

where, X_t:time series, t=1,2,3,L,N.

R_N: adjusted range statistic

$\overline{X_n}$: mean

σ : standard deviation.

Rescaled Range

The rescaled range statistic has been used extensively since its formulation by Mandelbrot. For a time series of n observations, $X_1, X_2, X_3, L, X_{n'}$ the rescaled range statistic, denoted by Q(n), is defined as

$$Q(n) = \frac{1}{\sigma(n)} \left[\max \sum_{t=1}^{k} (X_t - \overline{X_n}) - \min \sum_{t=1}^{k} (X_t - \overline{X_n}) \right]$$

(11)

$$\sigma(n) = \sqrt{\frac{\sum_{t=1}^{n} (X_t - \overline{X_n})}{n-1}}$$

(12)

where, X_t:time series, t=1,2,3,L,N.

N: size of the entire series

n: size of the partial series ($1 \leq n \leq N$)

σ (n): the standard deviation of the partial series.

[3], used so many annual time series to estimate the Hurst exponent and obtained the following relations with the calculation of Q(n)

$$\frac{Q(n)}{\sigma(n)} = \alpha n^{H}$$

(13)

where,H: Hurst exponent

n: size of the partial series ($1 \leq n \leq N$)

α: constant.

Equation (13) can be transformed with the logarithm and Hurst exponent can be estimated from the equation

$$\text{Log}(R(n)/\sigma(n)) = \text{Log}(\alpha) + H(\text{Log}(n))$$

(14)

Modified Rescaled Range

[13], pointed that the regression coefficients can be biased by the autocorrelation when the Hurst exponent is estimated by the regression equation from the known rescaled range. Thus, the rescaled range is a proper method for the time series of the long term memory but it is not for the short term memory. [13], developed the modified rescaled range method like the Equations (15) to (17) and [14,15] applied the method.

$$Q(n,q) = \frac{1}{\sigma_n(q)}\left[\max\sum_{t=1}^{k}\left(X_t - \overline{X_n}\right) - \min\sum_{t=1}^{k}\left(X_t - \overline{X_n}\right)\right]$$

(15)

$$\sigma_n(q) = \sigma_x^2(n) + 2\sum_{j-1}^{q} w_j(q)\gamma_j$$

(16)

$$w_j(q) = 1 - \frac{J}{q+1}, \quad q < n, \, j = 1,2,3,\cdots,n$$

(17)

where, $\sigma_n(q)$: weighted sum of autocovariances

$w_j(q)$: weighted autocovariance function

γ_j: autocovariance estimator

$\sigma_x^2(n)$: sample variance

q: Truncation lag of the weighted autocovariance function.

Here, we may carefully determine the truncation lag, q of weighted autocovariance function for the application of the modified rescaled range method. If q is so small, the effect of short range could not be considered, while q is so big the long range could be ignored [16]. Thus, the optimal truncation lag, q_{opt} should be estimated. [17], suggested the $q = n^{0.25}$ and [16] Equation (18). [15], used the modified rescaled range method with q = N/10. However this study will use the Equation (18) [7].

$$q_{opt} = \left(\frac{3N}{2}\right)^{1/3} \left(\frac{2\rho}{1-\rho^2}\right)^{2/3}$$

(18)

where, ρ: first order autocorrelation coefficient.

1/f Power Spectral Density Analysis

Power Spectral Density (PSD) has been used in the fields such as physiology, hear rate variability, and the self similarity of brain waves for the estimation of Hurst exponent. Trend in the PSD is proportional to $1/f^\alpha$ and the trend of 1/f is represented by the straight line with the slope of α and the $-\alpha$ can be calculated from the regression of low frequency band. The slope of α and Hurst exponent have the following relationship [6].

$$h(f) = \sigma^2 (2\pi)^{-1} \sum_{k=-\infty}^{+\infty} \rho(k) e^{ikf}$$

(19)

where, h(f): 1/f power spectral density

$$\alpha > 1 \rightarrow H = \frac{\alpha - 1}{2}$$

(20)

$$\alpha < 1 \rightarrow H = \frac{\alpha + 1}{2}$$

(21)

Detrended Fluctuations Analysis

The method of DFA has proven as a useful tool in revealing the extent of long-range correlation in time series. Firstly, the time series to be analyzed (with N samples) is integrated and the integrated time series is divided into boxes of equal length, n. In each box of length n, a least squares line is fit to the data (representing the trend in that box). The y-axis of the straight line segments is denoted, by $y_n(k)$. Next, we detrend the integrated time series, $y(k)$, by subtracting the local trend, $y_n(k)$, in each box. The root mean square fluctuation of this integrated and de-trended time series is calculated by Equations (22) and (23)

$$y(k) = \sum_{i=1}^{k} \left[X(i) - X_{\text{mean}} \right]$$

(22)

$$F(n) = \sqrt{\frac{1}{N} \sum_{k=1}^{N} \left[y(k) - y_n(k) \right]^2}$$

(23)

This computation is repeated over all time scales (box sizes) to characterize the relationship between F(n). The average fluctuation, as a function of boxes size, Typically, F(n) will increase with size n. A linear relationship on a log-log plot indicates the presence of power law scaling. Under such conditions, the fluctuations can be characterized by a scaling exponent d, the slope of the linear relation of log F(n) vs. logn.

The d has the following relationship and this relationship is represented by Equation (24). Then, The Hurst exponent is estimated by Equations (20) and (21).

$$\alpha = 2d - 1$$

(24)

Aggregated Variance Time Metho

Firstly, the mean value is obtained with the block of N/m by using Equations (25) and (26) for the time series X_i i=1,2,L,N (see the Figure 2).

$$X^{(m)}(k) = \frac{1}{m} \sum_{i=(k-1)m+1}^{km} X_i, \quad k = 1, 2, 3, \cdots, [N/m]$$

(25)

$$VarX^{(m)} = \frac{1}{N/m}\left(X^{(m)}(k) - \overline{X}\right)^2$$

(26)

For successive values of m, the sample variance of the aggregated series is plotted by m vs. log-log plot. The result should be a straight line with a slope of β and β has the following relationship:

$$H = 1 - \beta/2$$

(27)

In practical, the slope is estimated by fitting a least-squares line to the points of the plot.

Maximum Likehood Estimation (MLE)

Here the d is estimated by the S-MLE function of S-Plus [18], and the d is related to the Hurst exponent as follows;

$$d = H - 0.5$$

(28)

Applications of the Hurst Exponent Estimation Methods

Finally, complete content and organizational editing before formatting. Please take note of the following items when proofreading spelling and grammar: This section is to apply previous mentioned techniques for the estimation of Hurst exponent to the time series in a Section 2.1.Figure 3 shows the autocorrelation functions (ACF) of the time series. If we see the ACFs, the white noise and logistic map show the short term memory and others show the long term memory. Theoretical results for normal independent processes [3], indicated that asymptotically h = 1/2. One interpretation of the Hurst phenomenon has been to associate h = 1/2 with short memory models possessing short-term dependence structure, and h > 1/2 with long memory models possessing long-term dependence [19].

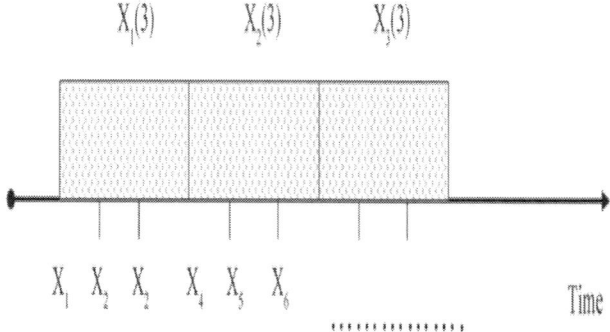

Figure 2: Aggregation of time series.

Rescaled Range and Modified Rescaled Range

The rescaled range and the modified rescaled range methods are applied to the time series for the error estimation of the methods. Figure 4 shows the comparison of Hurst exponents estimated by the rescaled and modified rescaled range methods. As we can see in Figure 4, the slopes estimated by the rescaled and modified rescaled range methods for the time series of short term memory characteristic are similar. However, the slopes for the time series of long term memory show relatively large differences.

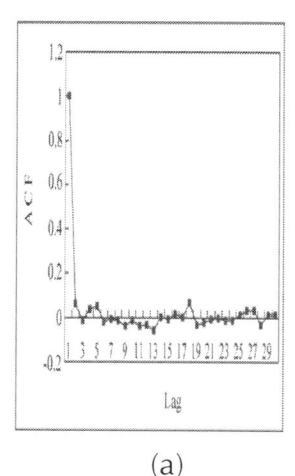

(a)

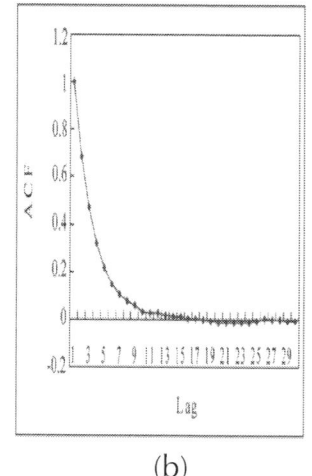

(b)

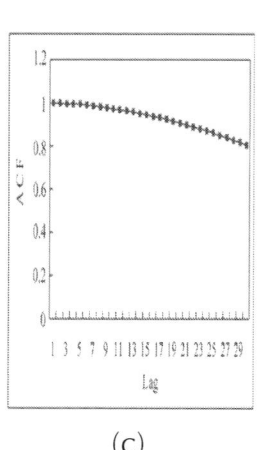

(c)

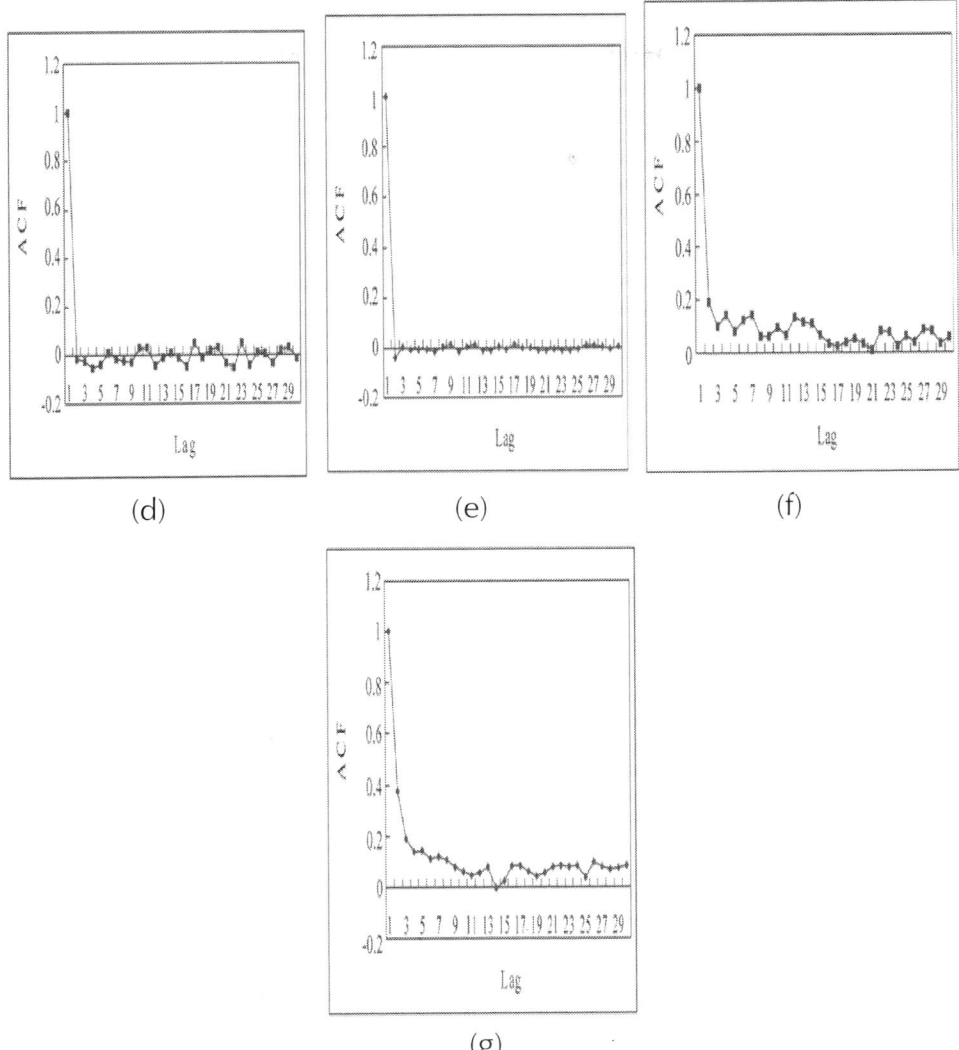

Figure 3: Autocorrelation functions of time series from chaotic and stochastic systems. (a) White noise; (b) AR(1); (c) Three torus; (d) Logisitic map; (e) TAR(2,1); (f) FARIMA; (g) FGN.

If the sample size is small, the error is small but the error is increased as the sample size is increased as shown in Figure 5. Even though the white noise is a random series, the Hurst exponent is estimated as 0.604 which represents the long term memory by the rescaled range method. However, if we use the modified rescaled range method the exponent is estimated as 0.55 which represents the short term memory (also see Figure 6).

1/f Power Spectral Density Analysis

We perform 1/f power spectral density (PSD or periodogram) analysis for the estimation of α and obtain the Hurst exponents by the Equations (20) and (21) for the time series mentioned in Section 2.1. Figure 7 shows the results of PSD analysis for each time series and this method shows relatively reasonable Hurst exponents as we can see in Table1 However, this PSD method may have a problem for the strong persistence system such as the three torus time series which is a quasi-periodic (see Table 1).

Detrended Fluctuation Analysis

This section estimates the Hurst exponents by using DFA method. The Hurst exponent is estimated by the Equations of (20) and (21). Table 2 shows the results of d and Hurst exponents. The DFA method shows the most reasonable Hurst exponent estimates for all-time series as shown in Table2

Aggregated Variance Time Method

In this section, we estimate Hurst exponent of time series by means of the aggregated variance time method (AVM). As shown in Figures 8 and 9, the AVM shows the property which the plotted points are scattered for the partial series is over a certain size, the slopes of the straight lines were obtained by the regression, and the Hurst exponents are estimated by Equation (27). However, the exponents are different from the known values. If we estimated the Hurst exponents after removing the scattered points the estimated exponents were similar with the known values (Tables 3 and 4).

Case I: Hurst exponent estimation by the regression with all points;

Case II: Hurst exponent estimation by the regression without the scattered points.

Maximum Likelihood Estimation (MLE)

We estimate the Hurst exponents for the time series by using the statistical package of S-MLE in S-Plus. The results are shown in Table 5 and we can know that the MLE method gives very reasonable values for each time series. Maximum likelihood estimation for FARIMA models can be performed in several ways [20]. We applied here an approximation in the spectral domain of the Gaussian maximum likelihood function, which was first proposed by [21], for short-memory models.

Results and Discussions

From the analysis of results, the Hurst exponent has estimated appropriately for the long term memory series by the adjusted and rescaled range methods but it was not for the short term memory like a white noise. The modified rescaled range method estimated the Hurst exponent properly for the short term memory series but it was not proper for the long term memory series. The 1/f PSD method estimated the Hurst exponents properly for the series except for the three torus case. The DFA and MLE methods have shown the reasonable results for the short and long term memory series. Especially, the DFA method is more convenient for the application than the MLE. Figure 6 shows the comparison for the Hurst exponent estimation results.

APPLICATION OF DFA FOR TREE-RING AND SOI SERIES

We found that the DFA is the most appropriate technique for the Hurst exponent estimation for both the short term memory and long term memory. In this section, we analyze 6 tree-ring series at USA sites and the SOI (Southern Oscillations Index) by means of DFA and the BDS statistic is used for nonlinearity test of the series. Especially, the BDS statistic is used for the nonlinearity of tree ring series tested by [14], who showed the tree ring series has random characteristics. The random characteristic means short term memory and may represent linear stochasticity of the series. Therefore we will examine the memory of tree ring series in the following sections.

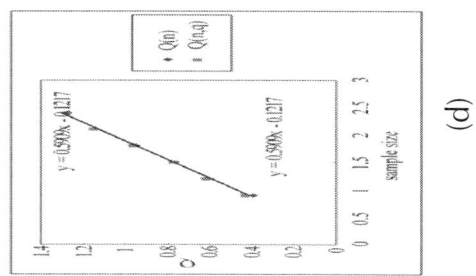

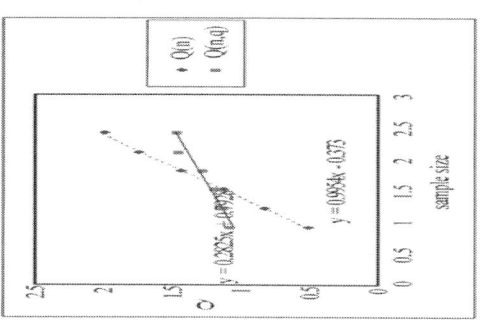

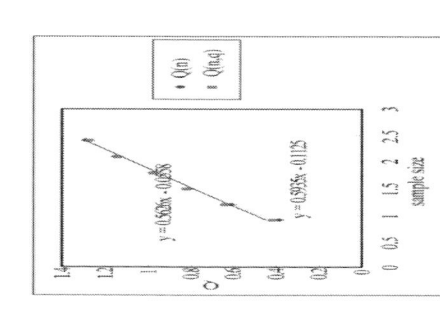

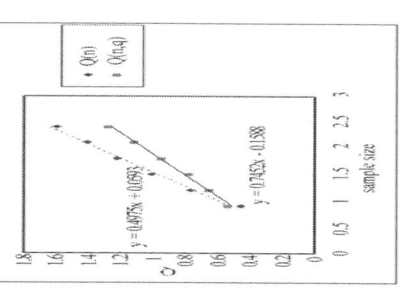

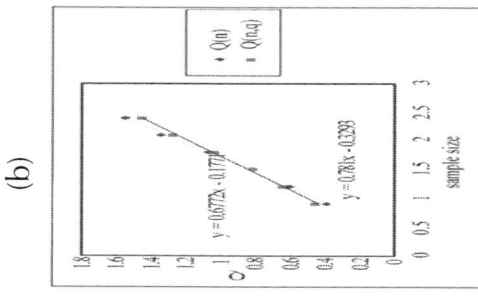

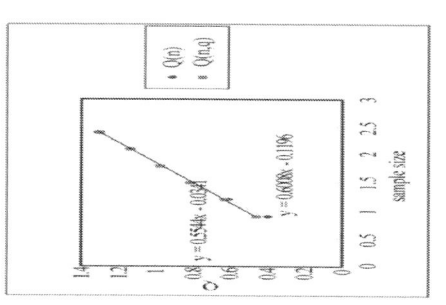

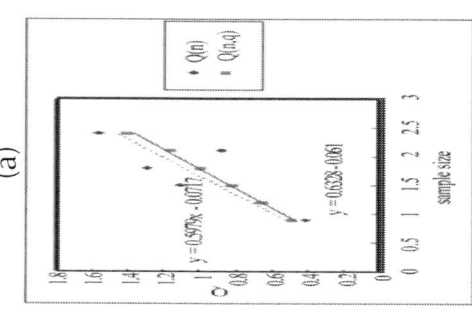

Descriptions of Tree-Ring and SOI Series

The climatic characteristics of an area are reflected in the growth of the trees. Consequently, the characteristics of tree are studied to determine the long-term climate behavior of a region. In this section, 6 tree ring series from USA are analyzed. The data sets of tree ring series consist of the period of 1892 to 1980 and sites are SACNDX, Spring, Freder, Calam, Hager, Dalton in California and Arizona. Tree-ring data are recorded as local climate properties and annual time scale. Each time series plot is shown in Figure 10.

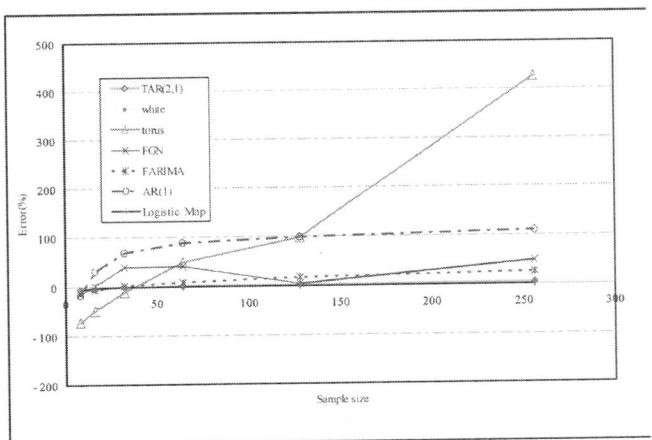

Figure 5: Errors for the rescaled range and modified rescaled range methods.

Figure 4: Estimation of Hurst exponent by rescaled range and modified rescaled range methods. (a) White noise; (b) AR(1); (c) Three torus; (d) Logistic map; (e) FGN; (f) FARIMA; (g) TAR(2,1).

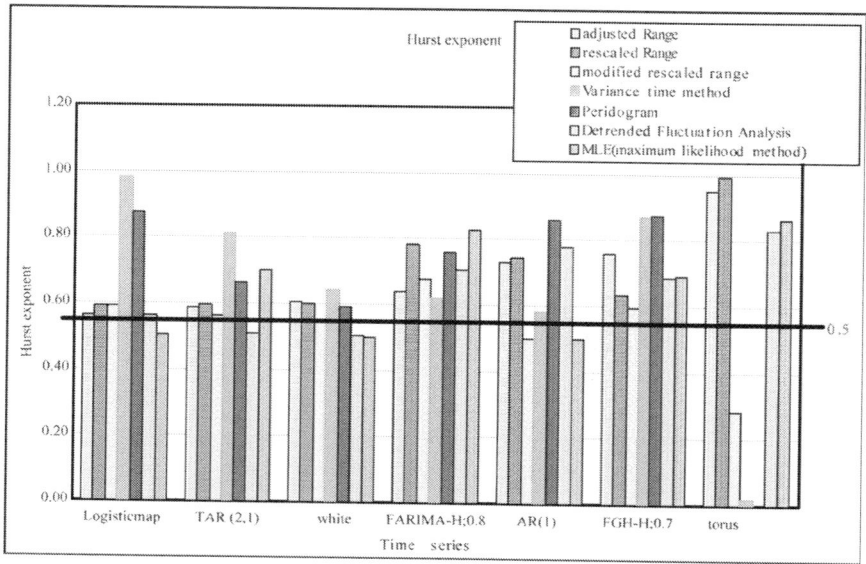

Figure 6: Comparison of the Hurst exponent estimates from various methods.

Table 1: 1/f spectral slope and hurst exponent

	White noise	AR(1)	Three torus	Logistic map	TAR(2,1)	FARIMA	FGN
α	0.180	0.716	5.287	0.745	0.323	0.519	0.746
H	0.590	0.858	∞	0.873	0.661	0.760	0.873

Table 2: d slope and hurst exponent

Data	Calam	Dalton	Freder	Hager	Sacndx	Spring	SOI
H	0.58	0.91	0.67	0.67	0.53	0.77	0.9

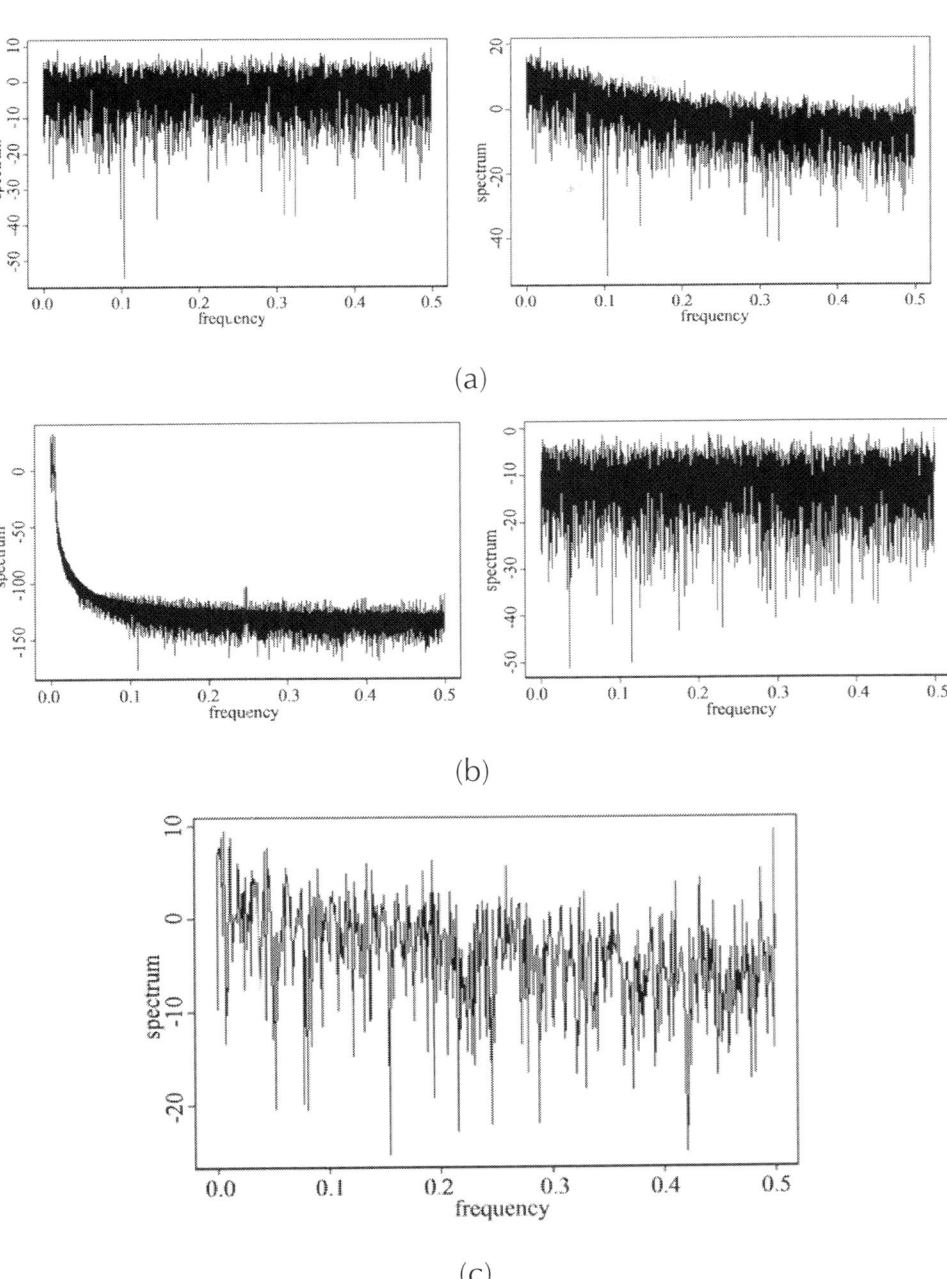

Figure 7: 1/f Power spectral density analysis. (a) White noise; (b) AR(1); (c) TAR(2,1); (d) FARIMA; (e) FGN.

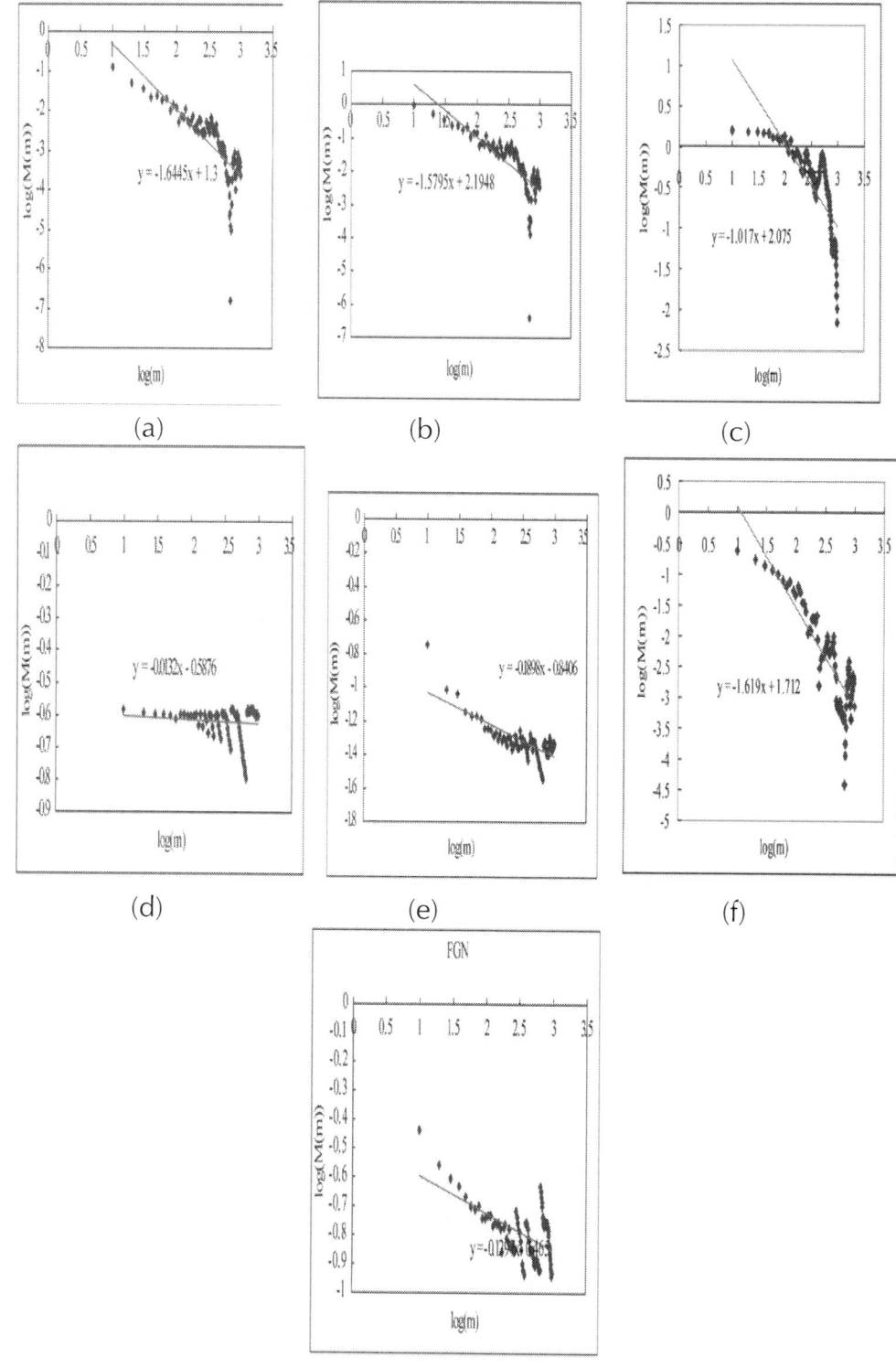

Figure 8: Estimation of Hurst exponent by Aggregated Variance Time method (case-I). (a) White noise; (b) AR(1); (c) Three torus; (d) Logistic map; (e) TAR(2,1); (f) FARIMA; (g) FGN.

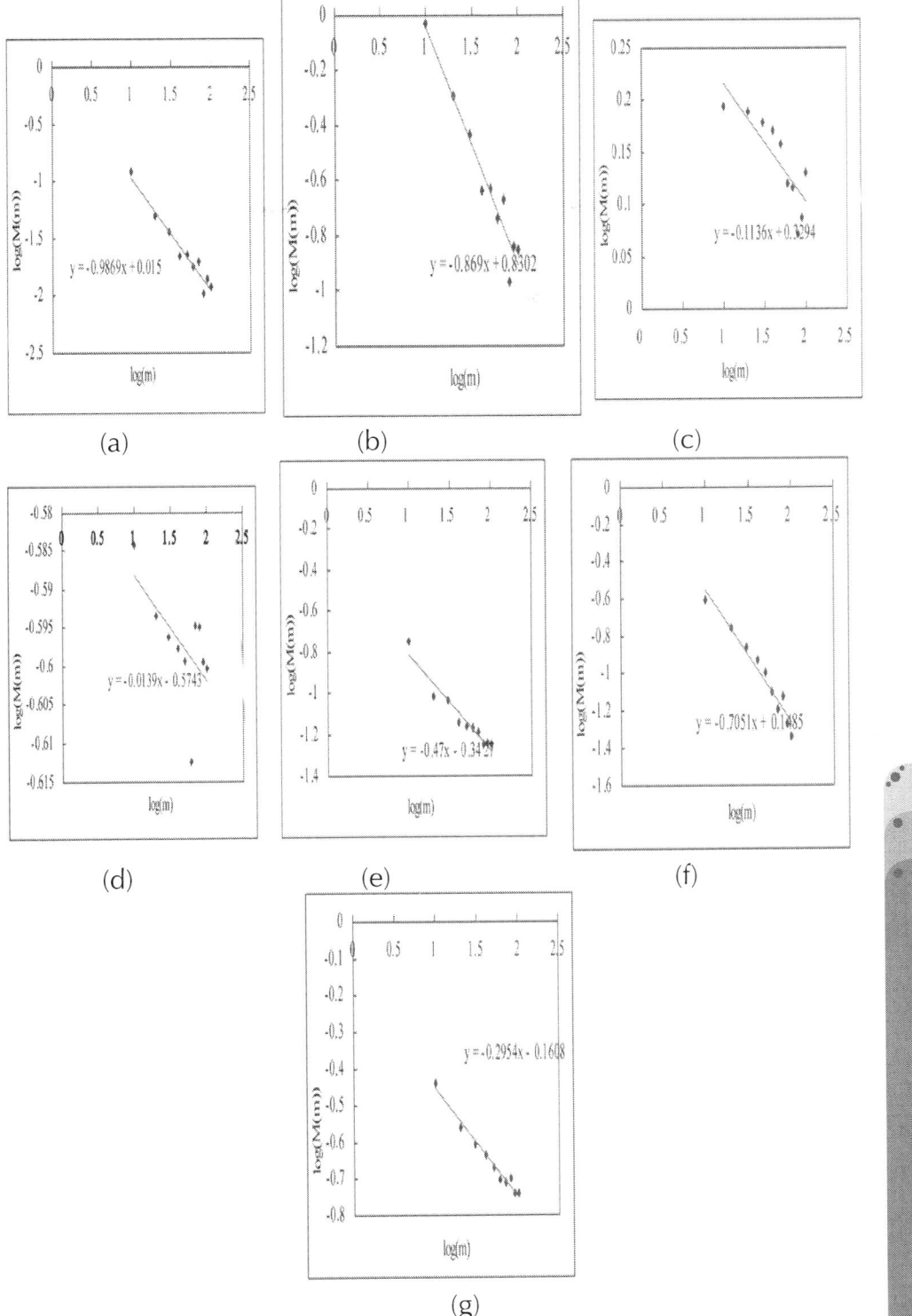

Figure 9: Estimation of Hurst exponent by aggregated variance time method (case-II). (a) White noise; (b) AR(1); (c) Three torus; (c) Three torus; (d) Logistic map; (f) FARIMA; (g) FGN.

Table 3: β slope and hurst exponent (case-I)

	White noise	AR(1)	Three torus	Logistic map	TAR(2,1)	FARIMA	FGN
β	1.645	1.580	1.017	0.013	0.189	1.619	0.129
Hurst expo-nent	0.170	0.210	0.491	0.934	0.905	0.190	0.930

$H = 1 - 2\beta$.

Table 4: β slope and hurst exponent (case-II)

	White noise	AR(1)	Three torus	Logis-tic map	TAR(2,1)	FARIMA	FGN
β	0.987	0.869	0.114	0.014	0.470	0.705	0.295
Hurst expo-nent	0.506	0.566	0.943	0.993	0.765	0.648	0.852

$H = 1 - 2\beta$.

Table 5: d slope and hurst exponent

	White noise	AR(1)	Three torus	Logis-tic map	TAR(2,1)	FARIMA	FGN
d	0.005	0.0004	0.364	0.023	0.199	0.326	0.295
Hurst expo-nent	0.5005	0.5004	0.864	0.502	0.699	0.826	0.852

The SOI is defined as the normalized pressure difference between Ta-hiti and Darwin. SOI values are calculated using the monthly mean sea level pressure (MSLP) data at papeete, Tahiti (149.6°W, 17.5°S) and Darwin, Australia (130.9°W, 12.4°S). In this study, we use the monthly SOI data from January of 1951 to December of 1999. Figure 11 shows the time series plot of SOI.

Hurst's Memory for Tree-Ring and SOI Data by DFA

Figure 12 shows the comparison of regression plots and results are shown in Table6 Figure 13shows the comparison of the autocorrelation (ACF) for tree-ring and SOI data. From the results, we found that the SOI series is time series which has a long term memory of H = 0.92. However, contrary to earlier work of [14], all the tree-ring series are not random and some of them show strong persistence from our analysis. A certain tree-ring series shows a long term memory of H = 0.97. Therefore, we can say that the tree ring and SOI series may show long term memory. And if the tree ring series are random as tested in [14], it may have linear stochasticity but the tree ring series which shows long term memory may have its nonlinearity and this could be modeled by the nonlinear stochastic models. Therefore, we would like to examine the nonlinearity of the tree ring and SOI series.

Nonlinearity for Tree-Ring and SOI Series

The BDS statistic is derived from the correlation integral and has its origins in the recent work on deterministic nonlinear dynamics and chaos theory. The method of delays can be used to embed a scalar time series $\{x_i\}, i=1,2,L,N$ into an m-dimensional space as follows

$$x_i = \left(x_i, x_{i+t}, \cdots, x_{i+(m-1)t} \right), \quad x_i \in R^m.$$

(29)

where t is the index lag. The correlation integral at embedding dimension m is given by

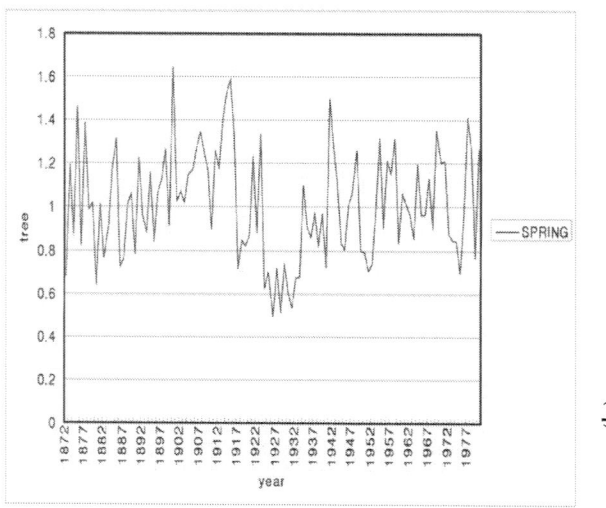

(b)

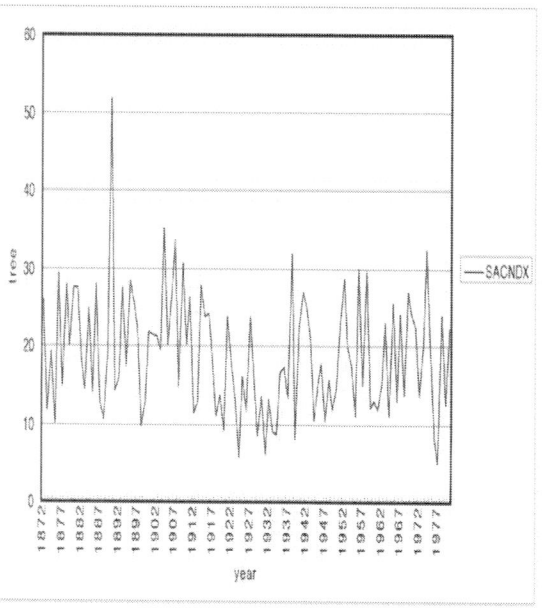

(a)

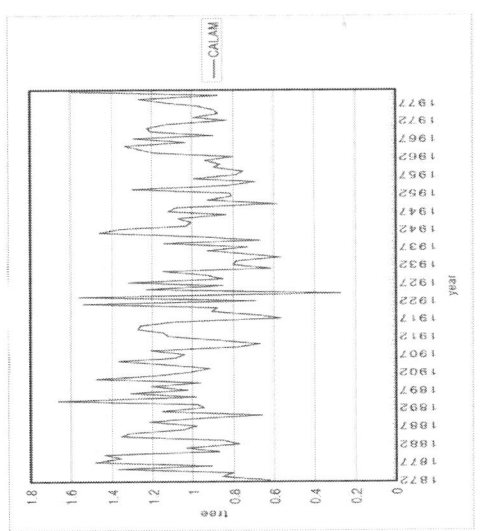

(d)

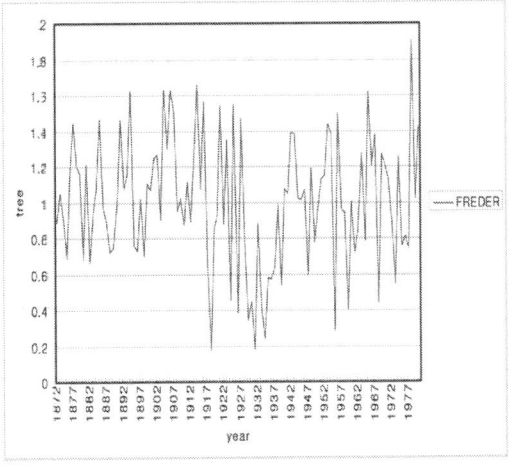

(c)

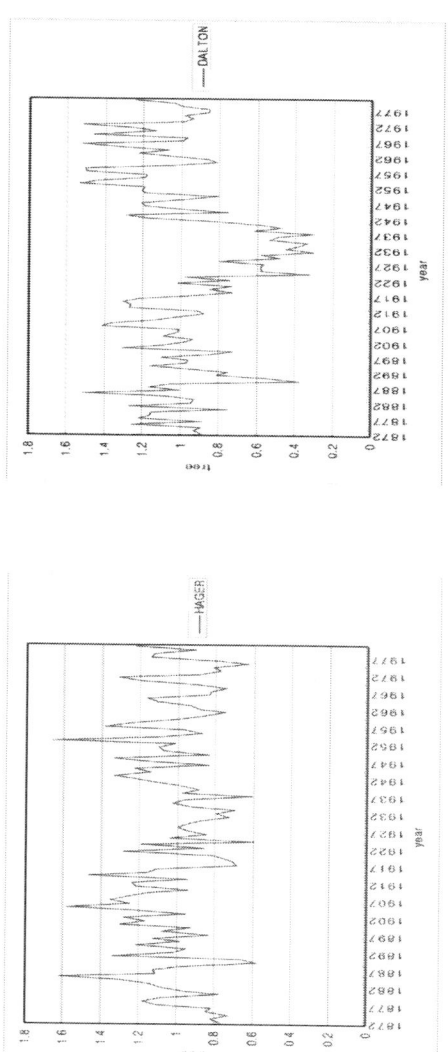

(e)

(f)

Figure 10: Time series plots of the tree-rings. (a) SACNDX; (b) Spring; (c) Freder; (d) Calam; (e) Hager; (f) Dalton.

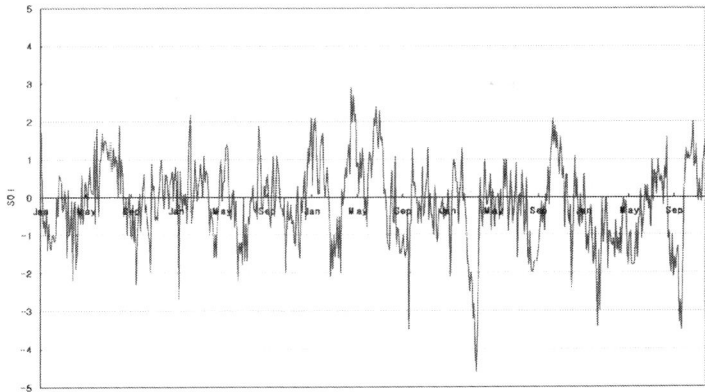

Figure 11: Time series plots of SOI.

Table 6: Estimations of the Hurst exponent by DFA for the tree rings and SOI series

Data	Calam	Dalton	Freder	Hager	Sacndx	Spring	SOI
H	0.58	0.91	0.67	0.67	0.53	0.77	0.9

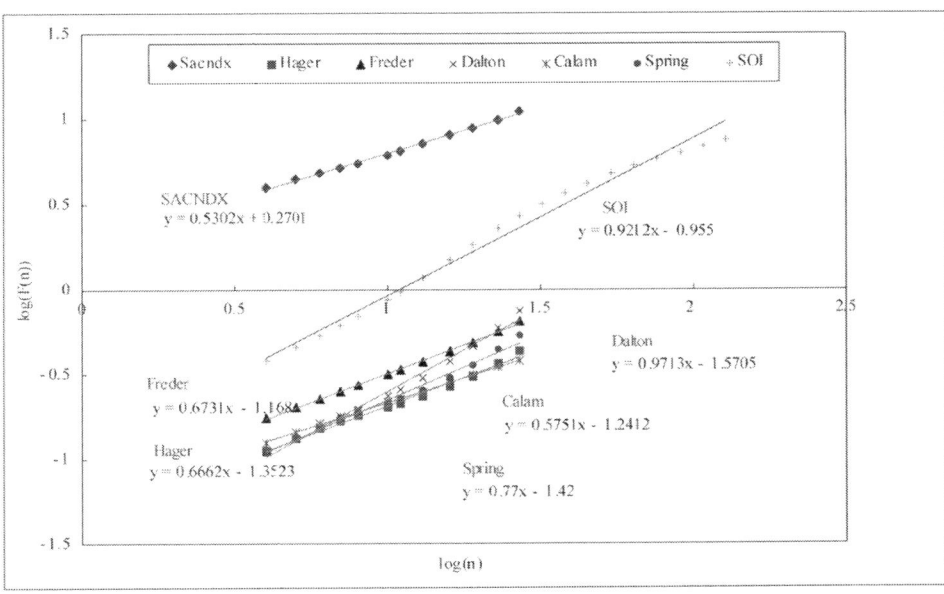

Figure 12: Comparison for regression plots by DFA.

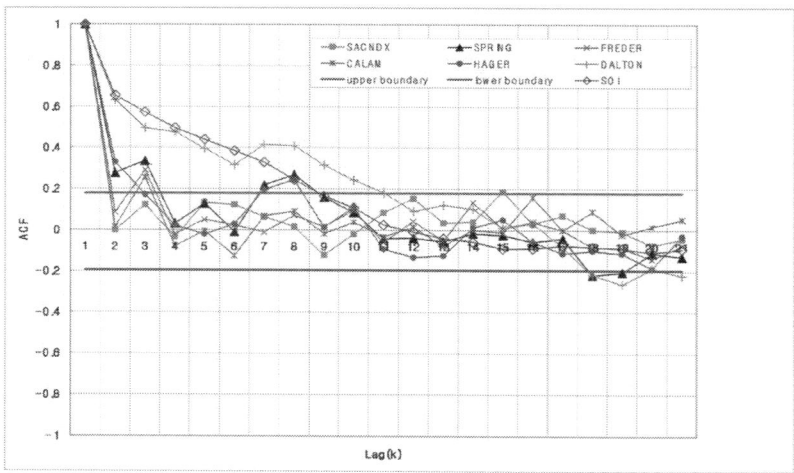

Figure 13: Comparison of the autocorrelation for tree-ring and SOI series.

$$C(m,N,r) = \frac{2}{M(M-1)} \sum_{1 \le i < j \le M} \Theta\left(r - \|x_i - x_j\|\right), \quad r > 0.$$

(30)

$$\Theta(a) = 0, \quad \text{if } a \le 0$$

$$\Theta(a) = 1, \quad \text{if } a \ge 0.$$

where N is the size of the data sets, M=N-(m-1) this the number of embedded points in m-dimensional space, and $\|\bullet\|$ denotes the sup-norm.C(m,N,r)measures the fraction of the pairs of points x_i,i=1,2,L,M, whose sup-norm separation is no greater than r. If the limit of C(m,N,r) as $N \to \infty$ exists for each r, we write the fraction of all state vector points that are within r of each other as

$$C(m,r) = \lim_{N \to \infty} C(m,N,r).$$

.

If the data is generated by a strictly stationary stochastic process which is absolutely regular, then this limit exists. In this case the limit is as follows

$$C(m,r) = \int\int_X \Theta(r - \|x - y\|)\,dF(x)\,dF(y), \quad r > 0.$$

(31)

When the process is IID, and since $\Theta(r - \|x - y\|) = \prod_{k=1}^{m} \Theta(r - |x_k - y_k|)$, Equation (31) implies that

C(m,r)=C^m(1,r). Also C(m,r)-C^m(1,r) has asymptotic normal distribution, with zero mean and variance as follows

$$\sigma^2(m, M, r)/4 = m(m-1)C^{2(m-1)}(K - C^2) + K^m - C^{2m}$$
$$+ 2\sum_{i=1}^{m-1}\left[C^{2i}\left(K^{m-i} - C^{2(m-i)}\right) - mC^{2(m-i)}\left(K - C^2\right)\right],$$

(32)

We can consistently estimate the constants C by C(1,r) and K by

$$K(m, M, r) = \frac{6}{M(M-1)(M-2)} \times \sum_{1 \le i < j \le M}\left[\Theta(r - \|x_i - x_j\|)\Theta(r - \|x_i - x_j\|)\right].$$

(33)

Under the IID hypothesis, the BDS statistic for m > 1 is defined as

$$\text{BDS}(m, M, r) = \frac{\sqrt{M}}{\sigma}\left[C(m, r) - C^m(1, r)\right].$$

.(34)

has a limiting standard normal distribution under the null hypothesis of IID as M ® ¥ and obtain its critical values using the standard normal distribution.

Before applying the BDS statistic, the first addressed issue is what region of "r" yields BDS statistic that are well approximated by the asymptotic distribution. As the sample size is increased, the distribution of the BDS statistic becomes more normal. So the minimal number of data must be provided. Next, the region of embedding dimension "m" should be suggested. If the sample size is fixed, we expect the finite sample property to worsen as "m" increases. This study follows the recommendation of [22] for selecting the ranges of m, r, N. Therefore, 500 or more observations are prepared and the embedding dimension m is

Table 7: The BDS statistics for tree-ring and SOI

r	Embedding dimension (m)	Tree-ring data						SOI
		Calam	Dalton	Freder	Hager	Sacndx	Spring	
0.5σ	2	−1.44(A)	8.07(R)	−1.08(A)	2.27(R)	−1.44(A)	3.01(R)	24.42(R)
	3	−1.05(A)	6.54(R)	−1.08(A)	2.48(R)	−2.24(R)	2.76(R)	26.75(R)
	4	−1.8(A)	5.21(R)	0.37(A)	0.67(A)	1.76(A)	1.61(A)	28.97(R)
	5	−2.64(R)	4.37(R)	3.68(R)	−3.73(R)	1.34(A)	0.98(A)	31.59(R)
1.0σ	2	1.6(A)	9.68(R)	0.07(A)	0.73(A)	−1.78(A)	3.19(R)	23.33(R)
	3	1.32(A)	9.07(R)	0.47(A)	0.67(A)	−2.18(R)	3.76(R)	25.09(R)
	4	1.03(A)	8.38(R)	1.67(A)	0.37(A)	−0.13(A)	3.42(R)	26.7(R)
	5	0.92(A)	8.61(R)	2.41(R)	0.49(A)	0.15(A)	3.99(R)	28.39(R)
1.5σ	2	1.7(A)	8.87(R)	0.74(A)	1.78(A)	−3.59(R)	3.5(R)	21.84(R)
	3	1.79(A)	8.7(R)	1.4(A)	1.03(A)	−3.12(R)	4.55(R)	22.86(R)
	4	1.55(A)	8.46(R)	2.2(R)	0.38(A)	−2.51(R)	4.56(R)	23.33(R)
	5	1.91(A)	8.56(R)	2.64(R)	0.51(A)	−2.37(R)	4.59(R)	23.61(R)
2.0σ	2	2.86(R)	9.86(R)	0.32(A)	0.93(A)	−3.11(R)	4.00(R)	21.37(R)
	3	2.87(R)	9.81(R)	0.91(A)	0.06(A)	−2.55(R)	4.97(R)	22.27(R)
	4	2.53(R)	9.54(R)	1.92(A)	−0.58(A)	−2.21(R)	4.78(R)	22.1(R)
	5	2.39(R)	9.52(R)	2.46(R)	−0.68(A)	−1.9(A)	4.56(R)	21.67(R)

Confidence interval: −1.96, +1.96 ($\alpha = 0.05$); A: accepted; R: rejected.

used in the range of 2 ≤m ≤5. Then, the value of "r" is selected as the half standard deviations of the data sets. The BDS statistic is a powerful tool for distinguishing random time series from the time series generated by nonlinear systems [22].

We use the BDS statistic for testing randomness of tree-ring and SOI data series and the results are shown in Table7 From the results, the SOI and certain tree-ring are representing their nonlinearities and thus could be modeled by the nonlinear type models [23].

SUMMARY AND CONCLUSIONS

This study has used the seven methods for the estimation of Hurst exponent and compared the results by applying each method to the generated time series form chaos and stochastic systems which have different characteristics. Then, this study discusses the advantages and disadvantages of the techniques and also the limitations of them.

1) The adjusted and rescaled range methods were proper for the long-term memory but the modified rescaled rang method was okay for the short-term memory. These methods have shown that the error was increased as the sample size was increased. Especially the error is increased for the series which the ACF is large.

2) The aggregated variance time method has shown that the plotted points on the proper scale were scattered. And so the Hurst exponent was appropriately estimated when we used the regression after removing the scattered points.

3) The 1/f PSD method has shown that it is a reasonable technique for the Hurst exponent estimation but it was not proper for a very strongly auto correlated series like a three tours system.

4) The DFA and MLE methods have shown that they are the most appropriate techniques for the Hurst exponent estimation for both the short term memory and long term memory.

From these results, we found that the DFA is the most appropriate technique for the Hurst exponent estimation for both the short term memory and long term memory. We analyzed the 6 tree-ring and SOI series for their memory tests by means of DFA and then the BDS statistic has used for nonlinearity test of the series. Our analysis has obtained the following conclusions 5) We found that SOI series is nonlinear time series which has a long term memory of H = 0.92. Contrary to earlier work of [14], all the tree-ring series are not random from our analysis. A certain tree ring series showed a long term memory of H = 0.97 and nonlinear property. Therefore, we can say that the SOI and tree-ring series may show long term memory and nonlinearity.

REFERENCES

1. H. Hurst, "Long-Term Storage Capacity of Reservoirs," Translation of the American Society of Civil Engineer, Vol. 116, 1951, pp. 770-799.
2. D. Koutsoyiannis and A. Efstratiadis, "Climate Change Certainty versus Climate Uncertainty and Inferences in Hydrological Studies and Water Resources Management," 1st General Assembly of the European Geosciences Union, Geophysical Research Abstracts, European Geosciences Union, Austria, Vol. 6, 2004.
3. B. Mandelbrot and J. Wallis, "Robustness of the Rescaled Range R/S in the Measurement of Noncyclic Long Run Statistical Dependence," Water Resources Research, Vol. 5, No. 5, 1969, pp. 967-988. http://dx.doi.org/10.1029/WR005i005p00967
4. P. Herman, "Physiological Time Series: Distinguishing Fractal Noises from Motions," European Journal of Physiology, Vol. 439, No. 4, 1999, pp. 403-415.
5. F. Pallikari, "Rescaled Range Analysis of Random Events," Journal of Scientific Exploration, Vol. 13, No. 1, 1999, pp. 25-40.
6. T. Bigger, C. Richard, A. B. Steinman, M. Linda, L. Joseph and J. Richard, "Power Law Behavior of RR-Interval Variability in Healthy Middle-Aged Persons, Patients with Recent Acute Myocardial Infarction, and Patients with Heart Transplants," Circulation, Vol. 93, No. 12, 1996, pp. 2142-2151. http://dx.doi.org/10.1161/01.CIR.93.12.2142
7. M. S. Taqqu and V. Teverovsky, "Estimation for Long Range Dependence: An Empirical Study," Fractals, Vol. 3, No. 4, 1995 pp. 785-798.http://dx.doi.org/10.1142/S0218348X95000692
8. Y. Zhao, "Self-Similarity in High Performance Network Analysis," University of Missouri-Columbia, Columbia, 1998.
9. S. Zafer and T. Sirin, "Traffic Engineering for Multimedia Networks; Data Collection on the Internet, Extensions to Wireless," NSF industry/University Co-Operative Research Center for Digital Video & Media, 1999.

10. M. A. Ausloos, "Statistical Physics in Foreign Exchange Currency and Stock Markets," Physica A, Vol. 285, No. 1-2, 2000, pp. 48-65. http://dx.doi.org/10.1016/S0378-4371(00)00271-5

11. J. W. Kantelhardt, B. E. Koscielny, H. A. Rego, S. Havlin and A. Bunde, "Detecting Long Range Correlations with Detrended Fluctuation Analysis," Physica A, Vol. 295, No. 3-4, 2001, pp. 441-454. http://dx.doi.org/10.1016/S0378-4371(01)00144-3

12. C. M. Kendziorski, "Evaluation Maximum Likelihood Estimation Methods to Determine the Hurst Coefficient," Physica A, Vol. 273, No. 3-4, 1999, pp. 439-451. http://dx.doi.org/10.1016/S0378-4371(99)00268-X

13. W. Lo, "Long Term Memory in Stock Market Prices," Economtrica, Vol. 59, No. 5, 1991, pp. 1279-1313. http://dx.doi.org/10.2307/2938368

14. R. Rao and D. Bhattacharya, "Hypothesis Testing for Long Term Memory in Hydrologic Series," Journal of Hydrology, Vol. 216, No. 3-4, 1999, pp. 183-196. http://dx.doi.org/10.1016/S0022-1694(99)00005-0

15. R. Rao and D. Bhattacharya, "Effect of Short-Term Memory on Hurst Phenomenon," Journal of Hydrologic Engineering, Vol. 6, No. 2, 2001, pp. 125-131. http://dx.doi.org/10.1061/(ASCE)1084-0699(2001)6:2(125)

16. V. Teverovsky, M. S. Taqqu and W. Willinger, "Acritical Look at Lo's Modified R/S Statistic," Journal of Statistical Planning and Inference, Vol. 80, No. 1-2, 1998, pp. 211-227.

17. P. Philips, "Time Series Regression with a Unit Roof," Econometrica, Vol. 55, No. 2, 1987, pp. 703-705. http://dx.doi.org/10.2307/1913237

18. MathSoft, "S-Plus2000; Guide to Statistical and Mathematical Analysis," StatSci Dvision, Seattle, 2000.

19. J. D. Salas and R. Pielke, "Stochastic Characteristics and Modeling of Hydroclimatic Processes," In: T. D. Potter and B. Colman, Eds., Handbook of Weather, Climate, and Water, Chapter 32, John Wiley & Sons, New York, 2002, pp. 585-603.

20. J. Beran, "Statistics for Long-Memory Processes," Chapman and Hall, New York, 1994.

21. P. Whittle, "Estimation and Information in Stationary Time Series," Arkiv för Matematik, Vol. 2, No. 5, 1953 pp. 423-434. http://dx.doi.org/10.1007/BF02590998

22. H. S. Kim, D. S. Kang and J. H. Kim, "The BDS Statistic and Residual Test," Stochastic Environmental Research and Risk Assessment, Vol. 17, No. 1-2, 2003, pp. 104-115.http://dx.doi.org/10.1007/s00477-002-0118-0

23. J. H. Ahn and H. S. Kim, "Nonlinear Modeling of El Nino/Southern Oscillation Index," Journal of Hydrologic EngineeringASCE, Vol. 10, No. 1, 2005, pp. 8-15. http://dx.doi.org/10.1061/(ASCE)1084-0699(2005)10:1(8)

CITATION

B. Kim, H. Kim and S. Min, "Hurst's Memory for Chaotic, Tree Ring, and SOI Series," Applied Mathematics, Vol. 5 No. 1, 2014, pp. 175-195. Doi: 10.4236/am.2014.51019.

Time-Domain Analysis of the Periodically Discontinuously Forced Fractional Oscillators

Zdzislaw Trzaska
Warsaw University of Ecology and
Management, Warsaw, Poland

ABSTRACT

A new method for the solution of non-sinusoidal periodic states in linear fractionally damped oscillators is presented. The oscillator is forced by a periodic discontinuous waveform and a viscous element is taken into account. The presented method avoids completely the Fourier series calculations of the input and output oscillator waveforms. In the proposed method, the steady-state response of fractionally damped oscillator is formulated directly in the time domain as a superposition of the zero-input and forced responses for each continuous piecewise segments of the forcing waveform, separately. The whole periodic response is reached by taking into account the continuity and periodicity conditions at instants of discontinuities of the excitation and then using the concatenation procedure for all segments. The method can be applied efficiently to discontinuous and continuous non-harmonic excitations equally well. Solutions are exact and there is no need to apply any of the widely up-to-date used frequency approaches. The Fourier series is completely cut out of the oscillator analysis.

INTRODUCTION

Mathematical models of dynamical systems with fractional-order derivatives have found many applications in various domains of science and technology such as viscoelasticity, control theory, electronics, heat conduction, electrode-electrolyte polarization, diffusion wave, electromagnetic waves, signal processing and many other physical processes [1] -[8] . In mechanics, for example, fractional-order derivatives have been successfully used to model the damping forces with memory effect or to describe state feedback controllers [9] [10]. Presently, it is clear that the fractional calculus broadens our perception not only of physical processes but also of many biological systems. Studies on dynamical behaviors of the electric signals of a human muscle's tremor of legs in a normal state and of the electric potentials of the human brain core from EEG's during epileptic seizure as well as a human hand finger tremor in Parkinson's disease have attracted considerable attention in many research centers through the world [11] [12] . These systems are known to display fractional-order dynamics (FOD). The characteristic features of all these models are that fractional derivatives introduce a new parameter-the order of fractional derivative n changing the properties of the solutions.

Recently, it is found in [13] that in fractional-order vibration systems of single degree of freedom, the term of fractional-order derivative whose order is between 0 and 2 acts always as damping force. In addition, almost all systems containing internal damping are not suitable to be described properly by the classical methods, but the fractional calculus represents one of the promising tools to incorporate in a single theory both conservative and non-conservative phenomena [14] . It is a well-recognized belief that fractional calculus leads to better results than classical one [15]. In some cases, it is possible to find out closed form solutions of fractional-order differential equations[2] [6] [8]. Therefore, a description of dynamical systems with using fractional derivatives may lead to results of major importance.

The definitions of fractional integral and derivative have been provided in the fractional calculus literature in a variety of ways, including Riemann-Liouville, Caputo, Erdélyi-Kober, Hadamard, Grünwald-Letnikov, and Riesz type. Equivalence of these definitions on some function has also been established [16] -[18] . However, the two most commonly used definitions are the Riemann-Liouville and the Caputo ones. It is well known presently that initial conditions are not taken into account in the same way whether Riemann–Liouville or Caputo definitions are considered. It is worth to mention that several mathematical and physical interpretations of fractional differentiation and of fractional systems exist in the literature [7] [14] . A demonstration of Montseny for a fractional integrator is now adapted to deduce a physical interpretation of a fractional system. From applications point of view the most convenient representation is that that permits to take into account initial condition in a coherent way with system physics.

Note that in the domain of fractional-order dynamical systems, fractional order harmonic oscillator is a fundamental issue, for it can be used to describe a much broader area of application than it is possible with the classical approach. Therefore, it is of fundamental importance to study the fractional model of the harmonic oscillator and to discuss the specific properties of its solutions.

The plan of the paper is the following: Section 2 provides basic formulations and introductory results to be used in this article. Section 3 derives the main problem concerning solutions of fractionally damped oscillators with periodic discontinuous excitations. In Section 4, we give results of applications of the proposed method to determination of T -periodic solution of equations describing fractionally damped oscillator with discontinuities in time periodic excitations. Finally, in Section 5 conclusions are presented.

BASIC FORMULATIONS AND INTRODUCTORY RESULTS

In this section we are focused on the equation of a linearly damped oscillator with the damping term generalized to a Caputo fractional derivative. We deal with the fractional-order differential equations ex-

pressed in terms of the Caputo derivatives needing the initial conditions in the same form as for the integer-order differential equations.
It is an advantage because applied problems require definitions of
fractional derivatives, where there are clear interpretations of initial
conditions, which contain $f(a), \dot{f}(a), \ddot{f}(a)$, etc. The analytic solution to
the fractionally damped equation can be drawn by means of Laplace
transform. Note, that the Laplace transform method is a very frequently
used tool for solving engineering problems.

The existence of periodic solutions is very often a desired property in
dynamical systems, constituting one of the most important research directions in applied mathematics, with applications ranging from celes-
tial mechanics to biology and finance. The analysis of linear harmonic
fractionally damped oscillator is a very recent and promising research
topic.

A standard approach to derive T-periodic solution is to define the differential operator which maps an initial value along the unique solution by T-units. Then the key periodicity and compactness conditions
are given such that some fixed point theorems can be applied to get
fixed points for the differential operator, which give rise to T-periodic
solutions.

In order to study periodic responses of a fractionally damped oscillator, we consider here the Caputo fractional derivative in the scalar
case, introduced in [14]. The considered oscillator is described by the
equation where $D = \frac{d}{dt}$ is the symbol of the derivative, $x(t) \in R$ and
$f_T(t) \in R$ denote the response and excitation terms of the oscillator,
respectively.

$$D^2 x(t) + AD^\nu x(t) + Bx(t) = f_T(t), \quad 0 \le t \le T$$

(1)

Constant coefficients A and B depend on parameters of the oscillator
elements and their connections. The fractional order n of the damping
term will be restricted to $0 \le n \le 1$.

The forcing term is assumed to be periodic discontinuous function of time t with finite number of discontinuities over period T. Using the concatenation procedure we represent the forcing term as splines of continuous segments (Figure 1).

For instance, the forcing term shown in Figure 1 can be represented as follows where $h = 0.5\left[1 + \mathrm{abs}(p - 0.5)/(p - 0.5)\right]$ denotes the concatenation factor with $p = p(t) = p(t + T)$ as the socalled saw tooth function [19].

$$f(t) = f(t + T) = \cos(0.8t) + h \cdot \left(1.2 - 0.4t - \cos(0.8t)\right) \tag{2}$$

It is worth pointing out that formula (2) can be easily extended on all remaining discontinuity points if they exist in the forcing terms. In the sequel we discus properties of fractionally damped oscillators with periodic non-sinusoidal discontinuous excitations and develop a systematic Fourier series-less method for their studies.

PRELIMINARY RESULTS

Analysis of Fractionally Damped Oscillators

In this section, the attention is focused on the linear fractionally damped oscillator characterized by the structure shown in Figure 2. A superconducting coil L with very small resistance R is connected in parallel with a supercapacitor C and controlled source as an active element. The forcing term is represented by independent current source $j_s(t)$.

Applying Kirchhoff current law and relations between current and voltage for each oscillator element yields where G denotes the active element conductance and μ and n denote fractional orders characterizing the super capacitor and superconducting coil, respectively.

$$LC\frac{\mathrm{d}^{\mu+\nu}x(t)}{\mathrm{d}t^{\mu+\nu}}+GL\frac{\mathrm{d}^{\nu}x(t)}{\mathrm{d}t^{\nu}}+x(t)=j_s(t)$$

(3)

Because in practice the sum of fractional orders is very near two thus in the sequel we take $\mu + \nu = 2$. Then, after few simple manipulations on the components of Equation (3) we can transform it to the form represented by (1) with In what follows the Equation (1) with notation (4) is considered taking into account the damping term generalized to a Caputo fractional derivative.

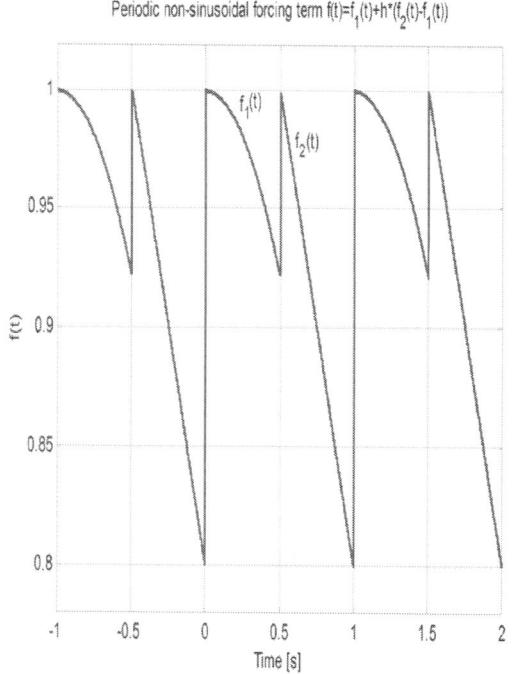

Figure 1: Periodic discontinuous forcing term f(t) = f(t + T).

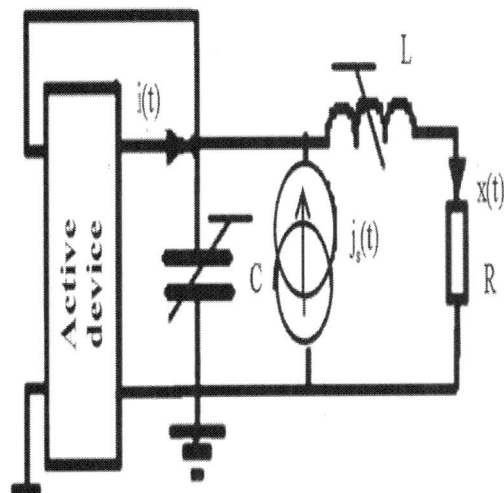

Figure 2: Scheme of the oscillator structure.

$A = G/C$, $B = 1/LC$ and $f_T(t) = (1/LC) j_s(t)$

(4)

It is worth mentioning that the Caputo derivative has been often used in fractional order systems since it has the practical initial states like that of integer order systems. Recall that the order of the derivative being considered is $0 \leq v \leq 1$. At the lower limit ($v = 0$) the equation represents a no damped oscillator and at the upper limit ($v = 1$) the ordinary linearly damped oscillator equation is recovered.

Time response of system (1) is thus defined by where $x_d(t)$ is the damped component, and $x_p(t)$ denotes the periodic component.

$x(t) = x_d(t) + x_p(t)$

(5)

In this article, we present a new method suitable for determination of non-sinusoidal periodic response $x_p(t)$ of linear fractionally damped oscillator. It avoids the Fourier series calculations of the input and output oscillator waveforms. In the proposed method, the oscillator response is formulated directly in the time domain as a superposition of the zero-input and forced responses for each continuous piecewise segments of the forcing term, separately. The strict periodic response is reached by taking into account the continuity and periodicity conditions at instants of discontinuities of the forcing term and then using the concatenation procedure for all segments.

It is hoped that a careful study of the analytic solution to the linear fractionally damped equation will help shed light on properties of the nonlinear equation and be of use for direct applications of fractionally damped oscillations.

To find an analytic solution we first consider homogeneous form of Equation (1) by annihilation of the forcing term. Thus, we take into considerations the equation in the following, we will present the solutions of this forcing free fractional differential equation for different fractional orders and will investigate the specific differences of the obtained solutions.

$$D^2 x_h(t) + AD^\nu x_h(t) + Bx_h(t) = 0, \quad 0 \le t < \infty \tag{6}$$

The analytic solution to the above fractionally damped equation is found by means of Laplace transform.

Using Caputo definition and applying Laplace transformation to both sides of (6) gives with $X_h(s) = L[x_h(t)]$ as the Laplace transform of the response of fractionally damped autonomous oscillator and x_0 and x_1 as initial conditions.

$$X_h(s) = \frac{\left(1 + As^{\nu-2}\right)sx_0 + x_1}{s^2 + As^\nu + B} \tag{7}$$

In order to evaluate the inverse Laplace transform of X(s) the following equation needs to be solved what is not a trivial problem for arbitrary n.

$$s^2 + As^v + B = 0 \tag{8}$$

Substituting $s = re^{ia}$ into (8) and then comparing to zero a real and imaginary part yields.

$$r^2 \cos(2\alpha) + Ar^v \cos(v\alpha) + B = 0,$$
$$r^2 \sin(2\alpha) + Ar^v \sin(v\alpha) = 0 \tag{9}$$

Performing detailed examinations of possible solutions of (9) it can be easily verified that for $0 £ n £ 1$ there are nine distinct cases as opposed to the usual three for the ordinary oscillator's equation (damped, over-damped, and critically damped). In three of these cases, the frequency of oscillation actually increases with increasing damping order before eventually falling to the limiting value given by the ordinary damped oscillator equation. For the six remaining cases, the behavior of the fractional oscillator is as expected and the frequency of oscillations decreases with increasing order of the fractional derivative. Observe moreover, that both terms of the second Equation in (9) would always be positive, thus, there are no solutions in the right half of the complex plane and none on the complete axis of negative real numbers. If there are solutions, they should be in pairs, complex conjugates, with $\pi/2 < \alpha < \pi$ and $-\pi/2 > \alpha > -\pi$. Thus, Equation (8) can be rewritten in

the form with $S_1 = r_1 e^{i\alpha_1}$ and $S_2 = r_1 e^{-i\alpha_1}$ where $r_1 > 0$ and $\pi/2 < \alpha_1 < \pi$ denotes solutions of (9).

$$(s - s_1)(s - s_2) = 0 \tag{10}$$

It has to be noted that for the fractionally damped equation repeated roots are not possible. The time response of system (6) is thus given by where C_1 and C_2 are constant. In Figure 3 are presented some graphs of solutions (11) for fixed A=0.5 and B =1.0, and various values of v.

$$x_h(t) = C_1 e^{s_1 t} + C_2 e^{s_2 t} \tag{11}$$

It should be emphasized that the evolution in time for this system is dominated by an exponential decay. Consequently, we obtain the result, that the behavior of the free solutions of the fractional harmonic oscillator under variation of the fractional derivative order n may be interpreted from a classical point of view as damping phenomena. Our view point sheds some new light on the arising question: does the oscillation frequency fall monotonically with respect to n? To demonstrate an answer we first examine the derivative of (8) with respect to n and get

$$\frac{ds}{d\nu} = -\frac{As^\nu \ln(s)}{2s + A\nu s^{\nu-1}}$$

(12)

Denoting s = a + iω and separating the imaginary part of (12) yields

$$\frac{d\omega}{d\nu} = -Imag\left(\frac{As^\nu \ln(s)}{2s + A\nu s^{\nu-1}}\right)$$

(13)

Considering this expression at ν = 0 we get initial slopes for the rate of change of ω with respect to ν as follows

$$\left.\frac{d\omega}{d\nu}\right|_{\nu=0} = -Imag\left(\frac{As^\nu \ln(s)}{2s + A\nu s^{\nu-1}}\right)\Bigg|_{\nu=0} = \frac{A\ln(A+B)}{4\sqrt{A+B}}$$

(14)

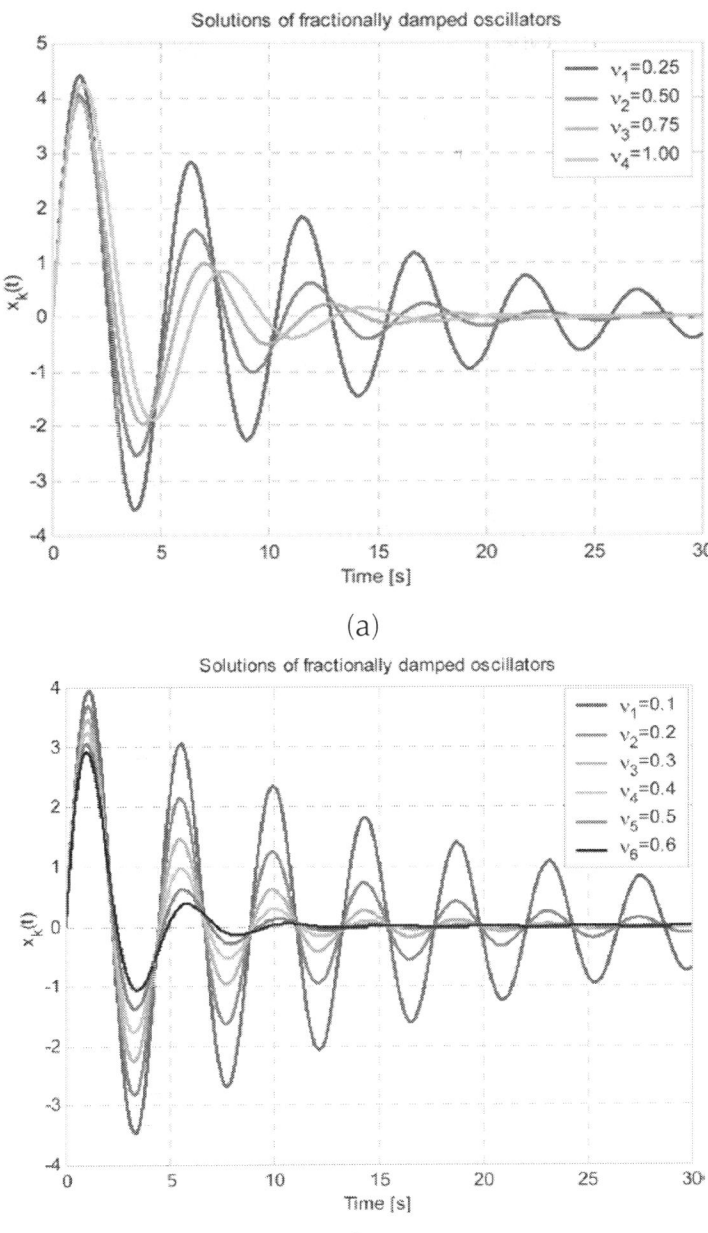

(a)

(b)

Figure 3: Solutions cf the fractionally damped oscillator with: (a) v = 0.25, 0.5, 0.27, 1.0; (b) v = 0.1, 0.2, 0.3, 0.4, 0.5, 0030.6.

Note that (14) generates three main cases of the initial slopes, namely

- The frequency initially increases with increasing damping order for A + B > 1;
- The frequency initially is not changing with increasing damping order for A + B = 1;
- The frequency initially decreases with increasing damping order for A + B < 1.

The above result indicates that there are nine cases for the linear fractionally damped oscillator because in each of the main cases can become any of the three particular cases, i.e., damped, over-damped, and critically damped. A surprising result lies in that that for three of the particular cases the oscillation frequency actually increases with increasing order of derivative of the damping term. After reaching a peak value, the frequency decreases as expected.

Fourier Series-Less Representation of Periodic Discontinuous Functions

In this subsection, we introduce notations, definitions, and preliminary facts, which are used throughout the remaining part of this paper. The representations introduced in the next section permits to take into account periodicity and continuity conditions of the permanent response of a fractionally damped oscillator in a coherent way with system physics. The main attention focuses on periodic discontinuous functions.

It is well known that a discontinuous function, like the square or sawtooth waveforms, cannot be expressed as a sum, even an infinite one, of continuous functions. The extraneous peaks in the square wave's Fourier series never disappear; because they occur whenever the function is discontinuous, and will always be present whenever the function has jumps [20] [21] . Quite obviously, if the excitation waveform is subject to jump changes the linear smoothing procedure is not a good choice anymore, because all conservative oscillator elements confuse and remove the high frequency components from the circuit output. For this reason, when a source waveform with jumps is applied to a linear oscillator it causes a typical effect of "edge blurring".

The main theorem concerning the convergence of the Fourier series at a discontinuity implies that this series converges to f(t) except at the point t = t_0, which is a point of discontinuity of f(t). Indeed, Gibbs [22] showed that if f(t) is piecewise smooth on [0, T] and t_0 is a point of discontinuity, then the Fourier partial sums will exhibit the same behavior, with the bump's height almost equal to Recall that the notations $f\left(t_0^+\right)$ and $f\left(t_0^-\right)$ represent the right-limit and left-limit, respectively, of f(t) at the point t_0.

$$\Delta f\left(t_0\right) = 0.18\left(f\left(t_0^+\right) - f\left(t_0^-\right)\right)$$

(15)

Thus, it is evident that for accurate analysis of fractionally damped oscillators excited by forcing terms producing complicated harmonic components, more formal time-domain mathematical tools are needed.

Taking into account the above requirements and insufficiencies of the methods based on Fourier series, which are up-to-date most commonly used for studies of periodic non-sinusoidal states of linear as well as nonlinear oscillators we propose in the sequel new method for obtaining, in closed form, the response of any linear oscillator corresponding to piecewise-continuous periodic non-sinusoidal forcing terms. The Fourier series-less method presented here depends on a "saw-tooth waveform" and of a scheme for the unified representation of composite periodic non-sinusoidal waveforms. It appears as a powerful broadly applicable technique that enables us to characterize non-harmonic periodic oscillations from a perspective different from that obtained by the method resulting from the Fourier series.

To avoid the difficulty appearing in practical applications of the Fourier series to obtain exact solutions of problems involved by periodic non-sinusoidal forcing terms operating in linear fractional order oscillators, we propose to use a so-called period carrying waveform p(t) defined as follows where T denotes the period.

$$p(t) = p(t+T) = \frac{T}{2} - \frac{T}{\pi} \operatorname{atan}\left(\cot\frac{\pi}{T}t\right)$$

(16)

The direct plot of (16) for T = π seconds is presented in Figure 4(a), and very often is called "saw-tooth function" or equivalently "saw-tooth waveform". Applying p(t) it is easy to generate a periodic form of the well-known absolute value function called also "vee" function which can be defined as follows

$$vee_T(t) = \sqrt{(p(t) - \tau)^2}$$

(17)

The plot of (17) for T = 2π seconds and τ = π seconds is presented in Figure 4(b).

In the sequel we will also take advantages of such useful functions as:

relay function r(t,τ) also called jump function which can be defined by

$$r(t,\tau) = \frac{Abs(t-\tau)}{t-\tau} = \frac{t-\tau}{Abs(t-\tau)}$$

(18)

The plot of function (18) depending on p(t), i.e. r(p(t),τ) for τ = π seconds is shown in Figure 5(a).

The switch-on function denoted by h(t,τ) is defined as follows

$$h(t,\tau) = \frac{1}{2}\left[1 + r(t,\tau)\right] = \begin{cases} 0 & \text{for } t \leq \tau \\ 1 & \text{for } t \geq \tau \end{cases}$$

(19)

The plot of function (19) depending on p(t), i.e. h(p(t),τ) for τ = π seconds is shown in Figure 5(b).

The above waveforms as well as many other similar ones, which can be easily derived on their base, are very useful in relatively simple representations of composite periodic non-sinusoidal waveforms. We remark that the latter situation will be relevant in the present analysis.

However, in a general setting, i.e., when we are dealing with waveforms x(t) exhibiting discontinuities, two restrictions must be fulfilled, namely: (i) global condition requiring that x(t) is absolutely integrable, (ii) local condition constraining x(t) to have a finite number of maxima and minima and a finite number of discontinuities in every finite interval [15] [22] . Moreover, we will use the results known in the general statement for the special case of the concatenation and show that this leads to an elegant procedure in Fourier series-less analysis also in the situation of real jumps in the input as well as output waveforms.

To cope with these effects we will describe discontinuous functions by using the saw-tooth waveform and its relatives such as switch-on and relay waveforms. To present the main idea of such an approach we consider at first a periodic non-sinusoidal function v(t) = v(t + T) with T = 0.5 seconds as the period. Such type of wave forms appears very often in power electronics [22] -[29] .

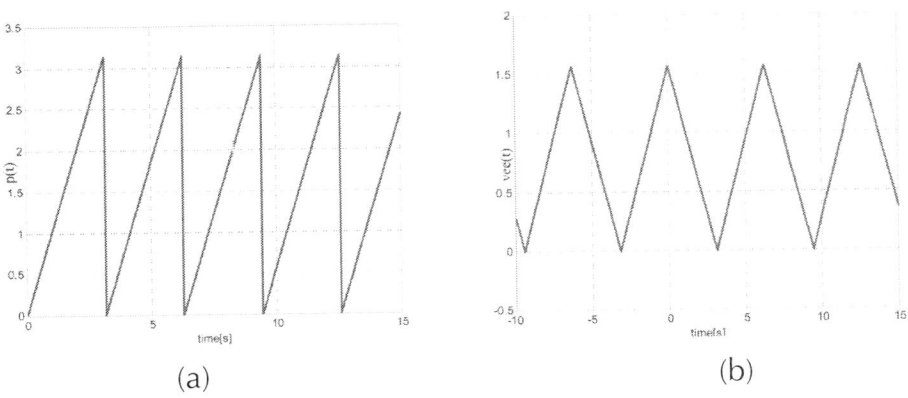

(a)   (b)

Figure 4: Diagrams of periodic functions: (a) p(t); (b) vee$_T$(t).

Its diagram is shown in Figure 6(a) and can be represented by with as the period carrying waveform.

$$v(t) = \cos\left(\frac{\pi}{4} p(t)\right)$$

$$(20)$$

$$p(t) = 0.25 - \left(\frac{0.5}{\pi}\right) \operatorname{atan}\left(\cot\left(\frac{\pi t}{0.5}\right)\right)$$

$$(21)$$

Further, the waveform f(t) = f(t + 1) shown in Figure 6(b) can be represented using the switch-on waveform h(t,τ) with τ = T/2 = 0.5 seconds as follows

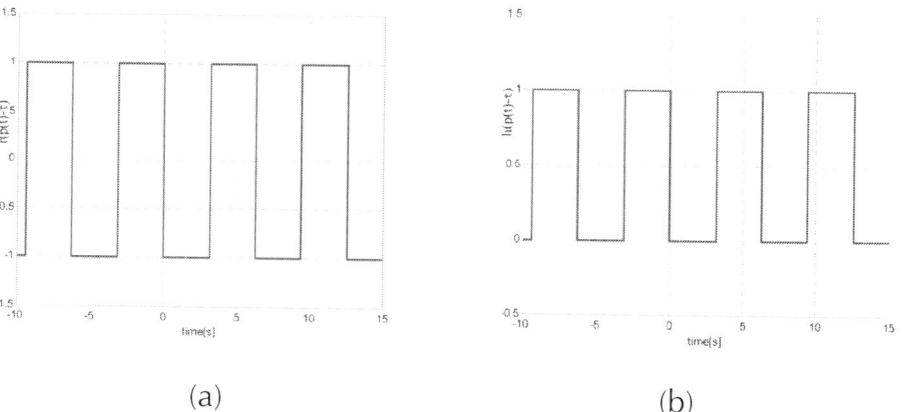

(a)   (b)

Figure 5: Diagrams of periodic discontinuous functions: (a) r(t); (b) h(t).

$$f(t) = f_1(p_1(t)) + h(p_1(t), T/2) \cdot \left[f_2(p_1(t)) - f_1(p_1(t)) \right]$$

$$(22)$$

where

$$f_1(t) = \cos(0.65t), \quad f_2(t) = 0; \quad p_1(t) = 0.5 - 1/\pi \cdot \operatorname{atan}\left(\cot(\pi t)\right)$$

$$(23)$$

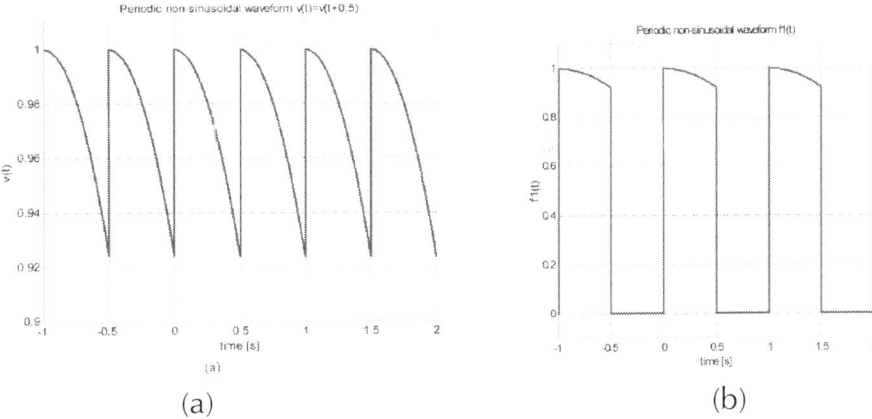

(a)   (b)

Figure 6: Periodic discontinuos functions: (a) $v(t) = \cos(\pi p(t)/4)$ and $T = 10.5$ s; (b) $f_1(t) = \cos(0.65t)$ for $t \in (0, 0.5)$ and $f_2(t) = 0$ for $t \in (0.5, 1)$ with T=1s.

The above examples show that the new proposed harmonic-less approach is very effective and leads to much less time consuming task than the standard method following from the Fourier series analysis. This intuitively appealing "switching rule" can be exploited in several ways. Using the switching approach (22) suggests the incorporation of a true smoothing element into the competition.

GENERAL ANALYSIS OF FORCED FRACTIONAL OSCILLATORS

Periodic Closed-Form Response

The purpose of this section lies in demonstrating how to generate and analyze complex form responses of linear systems under steady states with both discontinuous and continuous periodic non-harmonic forcing terms. As a matter of fact, periodic oscillations are very important and special phenomena not only in natural science but also in social science such as climate, food supplement, insecticide population, sustainable development [29].

The large applicability of the Caputo fractional derivatives is because we can formulate the fractional differential equations initial conditions as in the case of the classical one. Such particular properties of Caputo derivative as

$$D^v F = 0, \quad D^v t^p = \frac{\Gamma(p+1)}{\Gamma(p-v+1)} t^{p-v}, \quad D^v e^{\lambda t} = \lambda^n t^{n-v} E_{1,n-v+1}(\lambda t), \quad n-1 < v < n, \quad p > n-1$$

(24)

where F = const, $\Gamma(.)$ is the gamma function and $E_{1,\alpha}(.)$ denotes the two-parameter Mittag-Leffler function, are very useful in general analysis of forced fractional oscillators. The above features of the Caputo derivative have attracted the engineers' interest in the latter years, and now it is a tool used in almost every area of science.

To determine a periodic response of a fractionally damped oscillator we take into considerations the oscillator shown in Figure 1 in which source currents $j_s(t)$ represents the periodic discontinuous forcing term. Using notations (4) and approaching the time to the limit $t \to \infty$ we get

$$\lim_{t \to \infty} x_d(t) \to 0$$

(25)

so that

$$\lim_{t \to \infty} x(t) \to x_p(t) = x(t+T)$$

(26)

and in steady-state the oscillator response fulfills Equation (1) with x(t) = x(t + T) where T denotes the period of the forcing term $f_T(t) = Bj_s(t)$. Taking into account all particular continuous segments of the discontinuous forcing term we assume the corresponding continuous segment of the oscillator response as follows

$$x_k(t) = A_{1,k} e^{s_1 t} + A_{2,k} e^{s_2 t} + x_{f,k}(t), \quad k = 1, 2, \cdots, m$$

(27)

where s_1 and s_2 denote solutions of Equation (8), and $A_{1,k}$ and $A_{2,k}$ denote constants to be determined. The last component $x_{f,k}(t)$ takes the quite similar variation in time as that exhibited by the corresponding segment of $f_{T,k}(t)$ of the forcing term for $k = 1,2,\cdots,m$.

It has to be emphasized here that the oscillator is composed of a capacitor and an inductor both of fractional orders and they must fulfill the physical law of continuous changes of the capacitor voltage and inductor current. This important feature can be taken into account for determination of constants $A_{1,k}$ and $A_{2,k}$ of all segments of the oscillator response.

In the sequel to maintain only the clarity of the analysis without lost of its generality, we will limit the attention to the forcing current exhibiting time variations shown in Figure 7(a). This form of the forcing current is very often applied in practice for producing electrochemical nanostructural surface layers of appropriate metals or isolators to protect or improve exploitation properties of various construction and machinery details [8].

It is easily to check that the source current can be represented on the whole time interval by where $p(t)$ and $h(p(t), T_1)$ are determined by (16) and (19), respectively.

$$j_s(t) = J_1 + h\big(p(t), T_1\big) \cdot (J_2 - J_1) \tag{28}$$

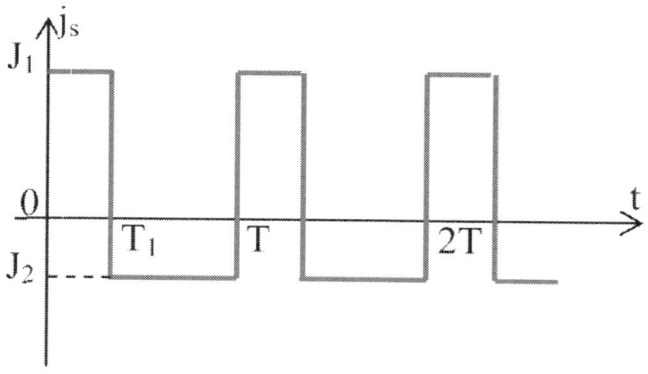

(a)

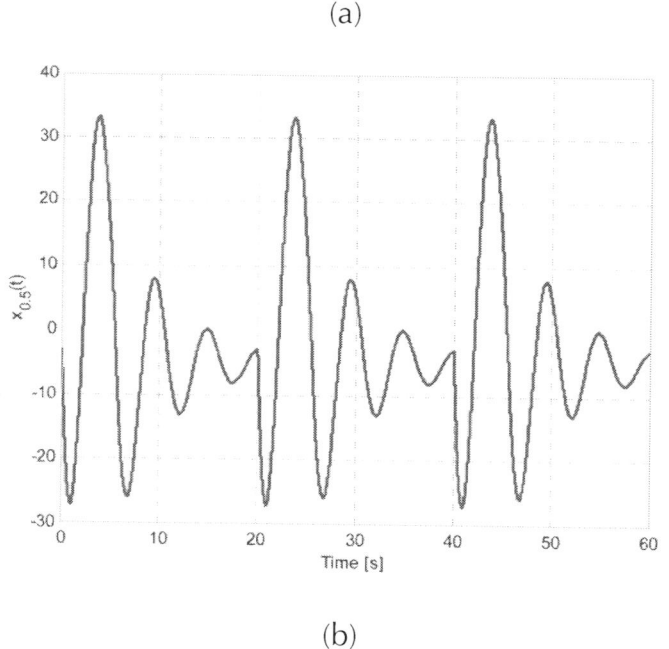

(b)

Figure 7: Diagrams of: (a) source current; (b) response of fractionally damped oscillator.

Note that in such a case the excitation current is characterized by two discontinuity moments, i.e. $t_1 = 0$ and $t_2 = T_1$ separating two segments of constant values J_1 and J_2, respectively. The response of the oscillator can be represented for $0 \leq t < T_1$ by

$$x_1(t) = A_1 e^{s_1 t} + A_2 e^{s_2 t} + X_1 \tag{29}$$

for $T_1 \leq t < T$ by

$$x_2(t) = A_3 e^{s_1 t} + A_4 e^{s_2 t} + X_2 \tag{30}$$

where constants A_1, A_2, A_3 and A_4 are to be determined. Components $X_1 = J_1$ and $X_2 = J_2$ correspond to particular solutions of Equation (1) for appropriate segments of the forcing current. The resulting current in the coil is expressed as follows

$$x(t) = x_1(t) + h(t, T_1) \cdot (x_2(t) - x_1(t)), \qquad 0 \leq t < T \tag{31}$$

In order to determine unambiguously the constants a help comes from the capacitor voltage, which in accordance with the principles of physics must be continuous in the time. So calculating the fractional derivative of the current in the coil, we get a quantity proportional to the capacitor voltage. Hence, taking into account the Caputo derivative for $0 \leq v \leq 1$ yields

$$D^v x_1(t) = s_1 t^{1-v} E_{1,2-v}(s_1 t) A_1 + s_2 t^{1-v} E_{1,2-v}(s_2 t) A_2,$$
$$D^v x_2(t) = s_1 t^{1-v} E_{1,2-v}(s_1 t) A_3 + s_2 t^{1-v} E_{1,2-v}(s_2 t) A_4 \tag{32}$$

Combining continuity and periodicity conditions for (29)-(32), we get a system of equations with unknown A_1, A_2, A_3 and A_4. In matrix notation, they take the following form

$$
\begin{bmatrix}
1 & 1 & -e^{s_1 T} & -e^{s_2 T} \\
e^{s_1 \bar{T}_1} & e^{s_2 \bar{T}_1} & -e^{s_1 \bar{T}_1} & -e^{s_2 \bar{T}_1} \\
0 & 0 & s_1 T_1^{1-v} E_{1,2-v}(s_1 T) & s_2 T_1^{1-v} E_{1,2-v}(s_2 T) \\
s_1 T_1^{1-v} E_{1,2-v}(s_1 T_1) & s_2 T_1^{1-v} E_{1,2-v}(s_2 T_1) & -s_1 T_1^{1-v} E_{1,2-v}(s_1 T_1) & -s_2 T_1^{1-v} E_{1,2-v}(s_2 T_1)
\end{bmatrix}
\begin{bmatrix}
A_1 \\ A_2 \\ A_3 \\ A_4
\end{bmatrix}
=
\begin{bmatrix}
X_2 - X_1 \\ X_2 - X_1 \\ 0 \\ 0
\end{bmatrix}
\tag{33}
$$

where T_1 denotes a moment at which exists a discontinuity of the forcing term $f_T(t)$. Solving the above equation for following parameters of the fractionally damped oscillator: $A = 0.5$, $B = 1$, $J_1 = 10$ A, $J_2 = -5$ A, $T = 20$ s and $T_1 = 5$ s and then substituting the result into (29)-(31) we get for $n = 0.5$ the coil current $x_{0.5}(t)$, which is presented in Figure 7(b).

Computer Simulations

In this subsection, the simulation results are presented to further demonstration of the reliability of the above approach. Three different fractional order systems represented by damped harmonic oscillator equation have been considered to confirm the given analytical results. First, let us use Equations (29)-(31) for $n = 0.25$, 0.5 and 0.75 for examinations of influences of the fractional order on the response damping within the period of forced fractional oscillator. Moreover, this gives also possibilities to examine the responses of the oscillator on different form of variations in time of the periodic forcing term.

Following the procedure presented in the above subsection, we can generate periodically forced responses of the same oscillator for different fractional orders. We exploit here a numerical technique using Matlab incorporated into the procedure, which is similar to the semi-analytical method. For forcing current exhibiting the discontinuous variations in time shown in Figure 7(a) with $J_1 = 10$A, $J_2 = -10$A, $T = 20$ s and $T_1 = 10$ s simulations were performed for oscillator parameters $A = 0.5$, $B = 1$ and fractional damping order: $v = 0.25$, 0.5 and 0.75, respectively. Results of numerical simulations are presented in Figure 8(a). Observe that the form of the periodic responses differ significantly with respect to that of the forcing term, but this fact agrees very well with physical considerations of the studied problem and the dynamic properties of the given oscillator. Moreover, there is no doubt that fractional order has become an exciting new factor indicating the damping rate of the oscillator. It takes quite similar role as the damping coefficient in the case of classic, i.e. integer order oscillator. When 3D phase trajectory is established (Figure 8(b)) in coordinates $(i_s(t), x_{0.5}(t),$

$\dot{x}_{0.5}(t)$ then it is evident that the oscillations generate a limit cycle with specific form being in good agreement to the physical state of the fractionally damped oscillator. From Figure 8(b), we can see that state

loops of $x_{0.5}(t)$ for $t \in (0, T_1)$ and $t \in (T_1, T)$ are similar to each other, but they are not identically coincided with each other although excitation current exhibits symmetry with respect to time axis. The difference is much more visible with changes of the fractional order n. Thus, it is expected that as in the case of classical oscillation systems, the form of the excitation term and parameters of the oscillator determine completely the evolution of the oscillation systems.

CONCLUSIONS

This work presents an effective concept of analyzing fractionally damped linear oscillator forced by a discontinuous periodic excitation and feasible computational strategies for time-varying dynamical processes based on continuity and periodicity approach for their solutions. This approach does not use any approximations based on the Fourier coefficient and not needs further investigation.

The relationship between fractional order and the damping of oscillations are discussed and concluded that the fractional order systems cannot be instead of by any other system. The stability criteria of fractionally damped oscillators are addressed. We have revealed that the fractionally damped oscillator contains nine subclasses of oscillators depending on parameters of elements and the fractional order n.

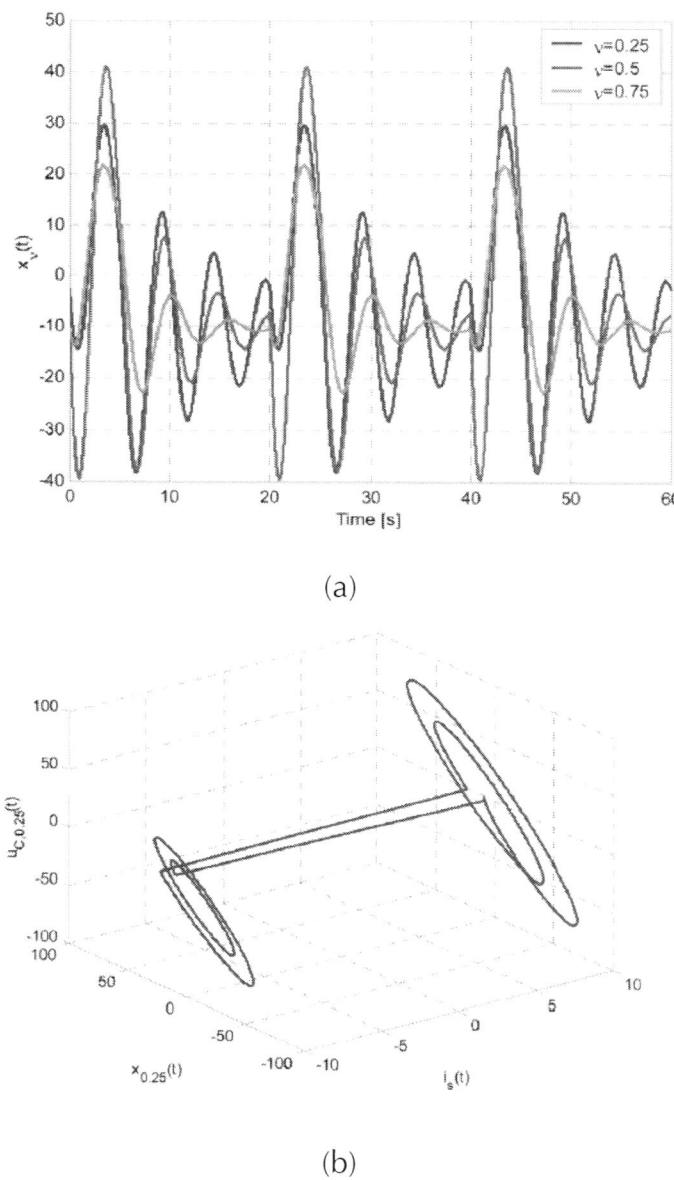

(a)

(b)

Figure 8: Results of numerical simulations: (a) oscillator responses for n = 0.25, 0.5, 0.75, (b) 3D plot for n = 0.25.

Methods are also proposed to obtain concatenation of continuous response segments corresponding to continuous segments of the discontinuous forcing term. A physical interpretation of a fractional order of the oscillator damping term is also proposed that demonstrates that any fractionally damped system can be viewed as an integer order system described by an equivalent diffusion term associated to a classical rational linear (exponentially damped) system.

We emphasize that it is possible to obtain exact periodic solutions for the output in periodic fractional-order dynamical systems forced by excitations being discontinuous in time.

REFERENCES

1. Magin, R.L. (2006) Fractional Calculus in Bioengineering. Begell House Publishers, Redding.
2. Monje, C.A., Chen. Y., Vinagre, B.M., Xue, D. and Feliu, V. (2010) Fractional-order Systems and Controls: Fundamentals and Applications. Springer, Berlin, New York.http://cx.doi.org/10.1007/978-1-84996-335-0
3. Chen, Y.Q., Ahn, H.S. and Podlubny, I. (2006) Robust Stability Check of Fractional Order Linear Time Invariant Systems with Interval Uncertainties. Signal Processing, 86, 2611-2618. http://dx.doi.org/10.1016/j.sigpro.2006.02.011
4. Carpinteri, A. and Mainardi, F. (1997) Fractional Calculus: Integral and Differential Equations of Fractional Order. Fractals and Fractional Calculus in Continuum Mechanics. Springer Verlag, Wien and New York, 223-276.http://dx.doi.org/10.1016/j.sigpro.2006.02.011
5. Lakshmikantham, V., Leela, S. and Devi, J.V. (2009) Theory of Fractional Dynamic Systems. Cambridge Academic Publishers, Cambridge.
6. Leela, S., Lakshmikantham, V. and Devi, J.V. (2012) Theory of Fractional Dynamic Systems. Cambridge Scientific Publishers, Cambridge.
7. Gutiérrez, R.E., Rosário, J.M. and Machado, J.T. (2010) Fractional Order Calculus: Basic Concepts and Engineering Applications. Mathematical Problems in Engineering, 2010, Article ID: 375858, 19 Pages.
8. Trzaska, M. and Trzaska, Z. (2011) Chaotic Oscillations in Fractional-Order Nonlinear Circuit Models of Bipolar Pulsed Electroplating, 20th European Conference on Circuit Theory and Design (ECCTD), Linkoping, 29-31 August 2011, 165-168.

9. Yulmetyev, R.M., Yulmetyeva, D. and Gafarov, F.M. (2005) How Chaosity and Randomness Control Human Health. Physica A, 354, 404-414.http://dx.doi.org/10.1016/j.physa.2005.02.036

10. Machado, J.A.T. (2002) Nonlinear Dynamics. An International Journal of Nonlinear Dynamics and Chaos in Engineering Systems. Special Issue of Fractional Order Calculus and Its Applications, 29.

11. Magin, R.L. and Ovadia, M. (2008) Modeling the Cardiac Tissue Electrode Interface Using Fractional Calculus. Journal of Vibration and Control, 14, 1431-1442. http://dx.doi.org/10.1177/1077546307087439

12. Sommacal, L., Melchior, P., Oustaloup, A., Cabelguen, J.-M. and Ijspeert, A.J. (2008) Fractional Multi-Models of the Frog Gastrocnemius Muscle. Journal of Vibration and Control, 14, 1415-1430. http://dx.doi.org/10.1177/1077546307087440

13. Herman, R. (2011) Fractional Calculus: An Introduction for Physicists. World Scientific & Imperial College Press, River Edge.

14. Podlubny, I. (1999) Fractional Differential Equations. Academic Press, San Diego.

15. Trzaska, Z. (2011) Matlab Solutions of Chaotic Fractional Order Circuits. In: Assi, A., Ed., Engineering Educations and Research Using MATLAB, Intech, Rijeka.

16. Lakshmikantham, V., Leela, S. and Devi, J.V. (2009) Theory of Fractional Dynamic Systems. Cambridge Scientific Publishers, Cambridge.

17. Belmekki, M., Nieto, J.J. and Rodriguez-López, R. (2009) Existence of Periodic Solution for a Nonlinear Fractional Differential Equation. Boundary Value Problems, 2009, 1-18.http://dx.doi.org/10.1155/2009/324561

18. Petras, I. (2011) Fractional-Order Nonlinear Systems, Modeling, Analysis and Simulation. Springer-Verlag, Berlin, Heidelberg. http://dx.doi.org/10.1007/978-3-642-18101-6

19. Trzaska, M. and Trzaska, Z. (2007) Straightforward Energetic Approach to Studies of the Corrosion Performance of Nanocopper Thin-Layers Coatings. Journal of Applied Electrochemistry, 37, 1009-1014.http://dx.doi.org/10.1007/s10800-007-9341-1

20. Mainardi, F. (1996) Fractional Relaxation-Oscillation and Fractional Diffusion-Wave Phenomena. Chaos, Solitons & Fractals, 7, 1461-1477. http://dx.doi.org/10.1016/0960-0779(95)00125-5

21. Cafagna, D. and Grassi, G. (2008) Fractional-Order Chua's Circuit: Time-Domain Analysis, Bifurcation, Chaotic Behavior and Test for Chaos. International Journal of Bifurcation and Chaos, 18, 615-639.http://dx.doi.org/10.1142/S0218127408020550

22. Jerri, A.J. (1998) The Gibbs Phenomenon in Fourier Analysis, Splines and Wavelet. Kluwer, Dordrecht. http://dx.doi.org/10.1007/978-1-4757-2847-7

23. Trzaska, Z. (2008) Fractional-Order Systems: Their Properties and Applications. Elektronika, 49, 137-144.

24. Trzaska, Z. (2012) Fractional-Order Harmonic Oscillators. Elektronika, 53, 162-167.

25. Trzaska, Z. (2010) Chaos in Fractional Order Circuits. Electrical Review, 86, 109-111.
26. Trzaska, Z. (2009) Fractional Order Model of Wien Bridge Oscillators Containing CPEs. Proceedings MATHMOD'09 Conference, Vienna, 357-361.
27. Luo, Y. and Chen, Y.Q. (2012) Fractional Order Motion Controls. John-Wiley and Sons Inc., New York. http://dx.doi.org/10.1002/9781118387726
28. Chen, Y.Q., Vinaigre, B.M., Xue, D. and Feliu, V.E. (2010) Fractional-Order Systems and Controls: Fundamentals and Applications. Springer, Berlin, New York.
29. Li, M., Lim, S.C. and Chen, S. (2011) Exact Solution of Impulse Response to a Class of Fractional Oscillators and Its Stability. Mathematical Problems in Engineering, 2011, Article ID 657839.

CITATION

Trzaska, Z. (2014) Time-Domain Analysis of the Periodically Discontinuously Forced Fractional Oscillators. Advances in Pure Mathematics, 4, 269-281. doi: 10.4236/apm.2014.46036.

Availability Equivalence Factors of a General Repairable Parallel-Series System

Abdelfattah Mustafa[1] and Ammar M. Sarhan[2]

[1]Department of Mathematics, Faculty of Science, Mansoura University, Mansoura, Egypt
[2]Department of Mathematics & Statistics, Dalhousie University, Nova Scotia, Canada

ABSTRACT

The availability equivalence factors of a general repairable parallel-series system are discussed in this paper considering the availability function of the system. The system components are assumed to be repairable and independent but not identical. The life and repair times of the system components are exponentially distributed with different parameters. Two types of availability equivalent factors of the system are derived. The results derived in this paper generalize those given in the literature. A numerical example is introduced to illustrate how the idea of this work can be applied.

INTRODUCTION

In reliability analysis, there are two main methods to improve non-repairable system design. These methods are the reduction and redundancy methods [1] . In the reduction method, it is assumed that the system design can be improved by reducing the failure rate(s) of a set of system components by a factor ρ, $0 < \rho < 1$, [1] -[4] . The redundancy method assumes that the system can be improved by increasing its components [5] .

There are more than one redundancy methods such as hot, warm, cold and cold with imperfect switch redundancy, named respectively as hot, warm, cold and cold with imperfect switch duplication methods [6] . The redundancy methods can be applied on repairable systems as well. In addition to the reduction method, the repairable system can be improved by increasing the repair rate of some of the system component(s) by a factors, $\sigma > 1$, [7] [8].

Using the redundancy method may not be a practical solution for a system in which the minimum size and weight are overriding considerations: for example, in satellites or other space applications, in well-logging equipment, and in pacemakers and similar biomedical applications [9] . In such applications space or weight limitations may indicate an increase in component performance rather than redundancy. Then more emphasis must be placed on better design, manufacturing quality control and on controlling the operating environment. Therefore, the concept of reliability/availability equivalence takes place. In such concept, the design of the system that is improved according to reduction or increasing method should be equivalent to the design of the system improved according to one of redundancy methods. That is, in this concept, one may say that the performance of a system can be improved through an alternative design [10] . In this case, different system designs should be comparable based on a performance characteristic such as 1) the reliability function or mean time to failure in the case of no repairs or 2) the availability in the case of repairable systems.

The concept of comparing different designs is applied in the literature in order to: 1) improve the reliability of a non-repairable system [11] ; 2) determine a representative service provider and create equivalent elements [12] ; 3) derive the reliability equivalence factors of some non-repairable systems [2] and the references therein; and 4) derive the availability equivalence factors of a repairable system [7] [8] .

The reliability equivalence concept applied on various non-repairable systems, [1] [2] [4] [13] -[17].

In this work, the reliability function and mean time to failure are used as characteristic measures to compare different system designs to derive reliability/mean time equivalence factors.

Repairable system indicates a system that can be repaired to operate normally in the event of any failure, such as automobiles, airplanes, computer network, manufacturing system, sewage systems, power plant or fire prevention system. Availability comprises "reliability" and "recovery part of unreliability after repair", indicating the probability that repairable systems, machines or components maintain the function at a specific moment [18] . It is generally expressed as the operable time over total time. Parallel-series system indicates sub-systems in which several components are connected in series, and then in parallel, or sub-systems that several components are connected in parallel, and then in series [19] . The reliability/availability of a parallel-series system has drawn continuous attention in both problem characteristics and solution methodologies[2] , [19] and [20] . Recently, [7] [8] discussed the availability equivalence factors of a repairable series-parallel system with independent and identical (non-identical) components.

Our goal in this paper is to derive the availability equivalence factors of a repairable parallel-series system with independent and non-identical components. The availability function of the system will be used as a performance measure to compare different system designs of the original system and other improved systems in order to derive these factors.

The structure of this paper is organized as follows. Section 2 introduces the illustration of the parallel-series system and the system availability. Section 3 presents the availabilities of the systems improved according to five different methods that can be applied to improve the performance of the original system. In Section 4, two types of availability equivalence factors of the system are discussed. A numerical example is introduced in Section 5 to illustrate how the idea of this work can be applied. Finally, Section 6 is devoted to the conclusions, which handle the main results derived throughout this work.

A GENERAL REPAIRABLE PARALLEL-SERIES SYSTEM

The system considered here consists of n subsystems connected in parallel, and with subsystem i consisting of m_i independent, repairable and nonidentical components connected in series for i=1,2,...,n We refer to such system as a general repairable parallel-series system. Figure 1 shows the diagram of that system.

Let T_{ij} and Y_{ij} be the lifetime and repair time, respectively, of component j in subsystem i, $1 \leq i \leq n$, $1 \leq j \leq m_i$. It is assumed that the life and repair times of component j in subsystem i, $1 \leq i \leq n$, $1 \leq j \leq m_i$, follow exponential distributions with failure rate λ_{ij} and repair rate μ_{ij}. Let N be the total number of the system components, that is $N = \sum_{i=1}^{n} m_i$.

Special Cases: This system generalizes the following cases:

1. Repairable parallel-series system with identical components, when $\lambda_{ij} = \lambda$, $\mu_{ij} = \mu$, j=1,2,...,m_i and i=1,2,...,n.
2. Repairable parallel system with non-identical components, when $m_i=1$ and i=1,2,...,n.
3. Repairable series system with non-identical components, when n=1 and j=1,2,...,m.

Let A_{ij}, be the availability of the component j in subsystem i and A_i be the availability of the subsystem i, $1 \leq i \leq n$, $1 \leq j \leq m_i$. One can easily derive A_{ij} and A_i, respectively, as, see [8]

$$A_{ij} = \frac{\mu_{ij}}{\mu_{ij} + \lambda_{ij}} = \frac{1}{1+\eta_{ij}}, \quad \text{where } \eta_{ij} = \frac{\lambda_{ij}}{\mu_{ij}},$$

(1)

and

$$A_i = \prod_{j=1}^{m_i} A_{ij} = \prod_{j=1}^{m_i} \left(\frac{1}{1+\eta_{ij}} \right).$$

(2)

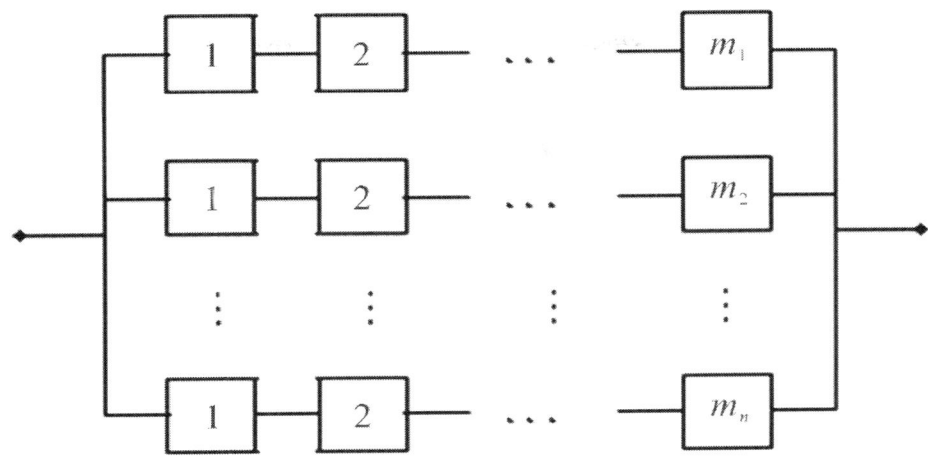

Figure 1: Parallel-series system structure.

Therefore, the system availability, denoted A_s, can be derived as

$$A_s = 1 - \prod_{i=1}^{n}(1 - A_i) = 1 - \prod_{i=1}^{n}\left[1 - \prod_{j=1}^{m_j}\left(\frac{1}{1 + \eta_{ij}}\right)\right].$$

(3)

DIFFERENT DESIGNS OF IMPROVED SYSTEM

The system can be improved according to one of the following three different methods:

1. Reduction method. In this method it is assumed that the component can be improved by reducing its failure rate by a factor ρ, $0 < \rho < 1$.
2. Increasing method. It is assumed in this method that the component can be improved by increasing its repair rate by a factor σ, $\sigma > 1$.
3. Standby redundancy method:
a. Hot duplication method: in this method we assume that the component is duplicated by an identical hot standby component.

b. Warm duplication method: in this method we assume that the component is duplicated by an identical warm standby component.
c. Cold duplication method: in this method we assume that the component is duplicated by an identical cold standby component.

In the following sections, we derive the availability of the system improved according to the methods mentioned above.

The Reduction Method

It is assumed in the reduction method that the system can be improved by reducing the failure rates of a set R components by a factor ρ, $0 < \rho < 1$. We assume that $R = \bigcup_{i=1}^{n} R_i$, where R_i is a set of the subsystem i components, $1 \leq i \leq n$. Also, we assume that $|R_i| = r_i$, $0 \leq r_i \leq m_i$

and

$$|R| = r = \sum_{i=1}^{n} r_i, \left(1 \leq r \leq N\right).$$

Let A_{ij}, ρ be the availability of the component j in subsystem i, improved by reducing its failure rate λ_{ij} by the factor ρ. One can easily derive

$$A_{ij,\rho} = \frac{1}{1 + \rho\eta_{ij}}, \text{ where } \eta_{ij} = \frac{\lambda_{ij}}{\mu_{ij}}.$$

$$(4)$$

Therefore, the availability of subsystem i improved by reducing the failure rates of a set R_i components by the factor ρ, denoted $A_{R_i,\rho}$, can be written as

$$A_{R_i,\rho} = \prod_{j \in R_i} A_{ij,\rho} \prod_{j \in \bar{R}_i} A_{ij} = \prod_{j \in R_i} \left(\frac{1}{1 + \rho\eta_{ij}}\right) \prod_{j \in \bar{R}_i} \left(\frac{1}{1 + \eta_{ij}}\right),$$

$$(5)$$

where $\bar{R}_i = M_i R_i$, M_i is the set of all subsystem i components, $M_i = \{1, 2, \ldots, m_i\}$, $1 \leq i \leq n$.

Finally, the availability of the system improved by reducing the failure rates of a set R components by the same factor ρ, denoted $A_{R,\rho}$, can be derived as

$$A_{R,\rho} = 1 - \prod_{i=1}^{n}\left[1 - \prod_{j \in R_i}\left(\frac{1}{1 + \rho\eta_{ij}}\right)\prod_{j \in \bar{R}_i}\left(\frac{1}{1 + \eta_{ij}}\right)\right].$$

(6)

The Increasing Method

It is assumed in the increasing method that the system can be improved by increasing the repair rates of a set S components by a factor σ, $\sigma > 1$. We assume that $S = \bigcup_{i=1}^{n} S_i$, where S_i is a set of the subsystem i components, $1 \leq i \leq n$. Also, we assume that $|S_i| = s_i, 0 \leq s_i \leq m_i$, and $|S| = s = \sum_{i=1}^{n} S_i$, $1 \leq s \leq N$.

Let $A_{ij,\sigma}$ be the availability of component j in subsystem i after increasing its repair rate μ_{ij} by the factor σ, $\sigma > 1$ and $A_{S_i,\sigma}$ be the availability of subsystem i which is improved by increasing the repair rates of a set S_i components by the same factor σ; and $A_{S,\sigma}$ be the availability of the system improved by increasing the repair rates of a set S components by the same factor σ. One can derive these availabilities in the following forms

$$A_{ij,\sigma} = \frac{\sigma\mu_{ij}}{\sigma\mu_{ij} + \lambda_{ij}} = \frac{\sigma}{\sigma + \eta_{ij}},$$

(7)

$$A_{S_i.\sigma} = \prod_{j \in S_i} A_{ij.\sigma} \prod_{j \in \overline{S_i}} A_{ij} = \prod_{j \in S_i} \left(\frac{\sigma}{\sigma + \eta_{ij}} \right) \prod_{j \in \overline{S_i}} \left(\frac{1}{1 + \eta_{ij}} \right),$$

(8)

$$A_{S.\sigma} = 1 - \prod_{i=1}^{n} \left[1 - \prod_{j \in S_i} \left(\frac{\sigma}{\sigma + \eta_{ij}} \right) \prod_{j \in \overline{S_i}} \left(\frac{1}{1 + \eta_{ij}} \right) \right],$$

(9)

Where $\overline{S_i} = MS_i$, for $1 \le i \le n$.

The Hot Duplication Method

It is assumed in the hot duplication method that the system can be improved by connecting every element in a set B components with an identical component in parallel. We assume that $B = \bigcup_{i=1}^{n} B_i$ where B_i is a set of the subsystem i components, $1 \le i \le n$ Also, we assume that $|B_i| = h_i$, $0 \le h_i \le m$, and $|B| = h = \sum_{i=1}^{n} h_i$, $1 \le h \le N$.

Let $A_{B_i}^H$ be the availability of the subsystem i which is improved by improving a set $B_i \subseteq M_i$ components, $1 \le i \le n$; and A_B^H be the availability of the system improved by improving a set B components according to the hot duplication method. One can derive

$$A_{B_i}^H = \prod_{j \in B_i} \left[1 - \left(1 - A_{ij} \right)^2 \right] \prod_{j \in \overline{B_i}} A_{ij} = \prod_{j \in B_i} \left[1 - \left(\frac{\eta_{ij}}{1 + \eta_{ij}} \right)^2 \right] \prod_{j \in \overline{B_i}} \left(\frac{1}{1 + \eta_{ij}} \right),$$

(10)

$$A_B^H = 1 - \prod_{i=1}^{n} \left[1 - \prod_{j \in B_i} \left[1 - \left(\frac{\eta_{ij}}{1 + \eta_{ij}} \right)^2 \right] \prod_{j \in \overline{B_i}} \left(\frac{1}{1 + \eta_{ij}} \right) \right],$$

(11)

where $\overline{B_i} = M_i B_i$ for $1 \le i \le n$.

The Warm Duplication Method

We say that, a component j in subsystem i is warm duplicated if it is connected in parallel with a non-identical component, having a failure rate v_{ij}, in parallel via a perfect switch. In the warm duplication method, it is assumed that the system can be improved when every component in a set B components is warm duplicated. We assume that $B = \bigcup_{i=1}^{n} B_i$, where B_i is a set of the subsystem i components, $1 \leq i \leq n$. Also, we assume that

$$|B_i| = w_i, \quad 0 \leq w_i \leq m_i.$$

and

$$|B| = w = \sum_{i=1}^{n} w_i, \quad 1 \leq w \leq N.$$

Let A_{ij}^{W} be the availability of the component j in the subsystem i when it is improved according to the warm duplication method. Using Markov process, A_{ij}^{W} can be obtained as follows, see [21] ,

$$A_{ij}^{W} = \frac{1 + \eta_{ij} + \xi_{ij}}{1 + \eta_{ij} + \xi_{ij} + \frac{1}{2}\eta_{ij}^2 + \frac{1}{2}\eta_{ij}\xi_{ij}},$$

$$(12)$$

where $\xi_{ij} = v_{ij} / \mu_{ij}$, for $1 \leq j \leq m_i$, and $1 \leq i \leq n$.

Let A_{Bi}^{W} be the availability of the subsystem i improved by improving B_i subsystem components according to the warm duplication method. Therefore, one can derive

$$A_{B_i}^{W} = \prod_{j \in B_i} \left(\frac{1 + \eta_{ij} + \xi_{ij}}{1 + \eta_{ij} + \xi_{ij} + \frac{1}{2}\eta_{ij}^2 + \frac{1}{2}\eta_{ij}\xi_{ij}} \right) \prod_{j \in \overline{B_i}} \left(\frac{1}{1 + \eta_{ij}} \right),$$

$$(13)$$

Finally, let A_B^W be the availability of the system improved by improving a set B components according to the warm duplication methods. Using Equation (13), we get

$$A_B^W = 1 - \prod_{i=1}^{n}\left[1 - \prod_{j \in B_i}\left(\frac{1 + \eta_{ij} + \xi_{ij}}{1 + \eta_{ij} + \xi_{ij} + \frac{1}{2}\eta_{ij}^2 + \frac{1}{2}\eta_{ij}\xi_{ij}}\right)\prod_{j \in \bar{B}_i}\left(\frac{1}{1 + \eta_{ij}}\right)\right].$$

(14)

The Cold Duplication Method

It is assumed in the cold duplication method, that each component of set B components is connected in parallel with an identical component via a perfect switch. We assume that $B = \bigcup_{i=1}^{n} B_i$, where B_i is a set of the subsystem i components, $1 \leq i \leq n$. Also, we assume that $|B_i| = c_i$, $0 \leq c_i \leq m_i$

and

$$|B| = c = \sum_{i=1}^{n} c_i, \, 1 \leq c \leq N.$$

Let A_{ij}^C is the availability of the component j in subsystem i when it is improved according to the cold duplication method; A_B^C be the availability of subsystem i, which is improved according to cold duplication method; and A_B^C be the availability of the system improved by improving set B components according to the cold duplication method. Using Markov process theory, A_{ij}^C is, see[22] ,

$$A_{ij}^C = \frac{\mu_{ij}^2 + \lambda_{ij}\mu_{ij}}{\mu_{ij}^2 + \lambda_{ij}\mu_{ij} + \frac{1}{2}\lambda_{ij}^2} = \frac{1 + \eta_{ij}}{1 + \eta_{ij} + \frac{1}{2}\eta_{ij}^2}.$$

(15)

Using Equation (15) and the nature of the series subsystem i, one can derive

$$A_{B_i}^C = \prod_{j \in B_i} \left(\frac{1 + \eta_{ij}}{1 + \eta_{ij} + \frac{1}{2}\eta_{ij}^2} \right) \prod_{j \in \bar{B}_i} \left(\frac{1}{1 + \eta_{ij}} \right).$$

(16)

Finally, using Equation (16) and the nature of the parallel connection of the subsystems, we get

$$A_B^C = 1 - \prod_{i=1}^{n} \left[1 - \prod_{j \in \bar{B}_i} \left(\frac{1 + \eta_{ij}}{1 + \eta_{ij} + \frac{1}{2}\eta_{ij}^2} \right) \prod_{j \in \bar{B}_i} \left(\frac{1}{1 + \eta_{ij}} \right) \right].$$

(17)

AVAILABILITY EQUIVALENCE FACTORS

In this section, we derive the availability equivalence factors of a repairable parallel-series system with independent, non-identical and repairable components. Two types of availability equivalence factors will be discussed. These two types are referred as availability equivalent reducing factor and availability equivalent increasing factor. Following the definition of reliability equivalence factors introduced in [1] .

Availability Equivalence Reducing Factor

Availability equivalence reducing factor, in short AERF, referred as $\rho = \rho_{R,B}^D$, $D = H$, W, C for hot, warm and cold, respectively, is defined as the factor ρ by which the failure rate of a set R components should be reduced in order to get equality of the availability of another better design which can be obtained from the original system by assuming hot, warm and cold duplications of a set B components. That is,

$\rho = \rho^D_{R,B}$, for D = H, W, C, is the solution of the following equations in ρ,

$$A_{R,\rho} = A^D_B, \quad D = H,W,C. \tag{18}$$

In what follows, we give the non-linear equations needed to be solved to get the three possible AERF's.

1) Hot availability equivalence reducing factor (HAERF): Substituting Equations (6) and (11) into Equation (18), $\rho = \rho^H_{R,B}$, is the solution of the following non-linear equation in ρ,

$$\prod_{i=1}^{n}\left[1-\prod_{j\in R_i}\left(\frac{1}{1+\rho\eta_{ij}}\right)\prod_{j\in \overline{R}_i}\left(\frac{1}{1+\eta_{ij}}\right)\right] = \prod_{i=1}^{n}\left[1-\prod_{j\in B_i}\left[1-\left(\frac{\eta_{ij}}{1+\eta_{ij}}\right)^2\right]\prod_{j\in \overline{B}_i}\left(\frac{1}{1+\eta_{ij}}\right)\right]. \tag{19}$$

2) Warm availability equivalence reducing factor (WAERF): Substituting Equations (6) and (14) into Equation (18), $\rho = \rho^W_{R,B}$, is the solution of the following non-linear equation in ρ,

$$\prod_{i=1}^{n}\left[1-\prod_{j\in R_i}\left(\frac{1}{1+\rho\eta_{ij}}\right)\prod_{j\in \overline{R}_i}\left(\frac{1}{1+\eta_{ij}}\right)\right] = \prod_{i=1}^{n}\left[1-\prod_{j\in B_i}\left(\frac{1+\eta_{ij}+\xi_{ij}}{1+\eta_{ij}+\xi_{ij}+\frac{1}{2}\eta_{ij}^2+\frac{1}{2}\eta_{ij}\xi_{ij}}\right)\prod_{j\in \overline{B}_i}\left(\frac{1}{1+\eta_{ij}}\right)\right]. \tag{20}$$

3) Cold availability equivalence reducing factor (CAERF): Substituting Equations (6) and (17) into Equation (18), $\rho = \rho^C_{R,B}$, satisfies the following non-linear equation

$$\prod_{i=1}^{n}\left[1-\prod_{j\in R_i}\left(\frac{1}{1+\rho\eta_{ij}}\right)\prod_{j\in \overline{R}_i}\left(\frac{1}{1+\eta_{ij}}\right)\right] = \prod_{i=1}^{n}\left[1-\prod_{j\in B_i}\left(\frac{1+\eta_{ij}}{1+\eta_{ij}+\frac{1}{2}\eta_{ij}^2}\right)\prod_{j\in \overline{B}_i}\left(\frac{1}{1+\eta_{ij}}\right)\right]. \tag{21}$$

Equations (19)-(21) have no closed solutions, therefore, a numerical technique method is needed to get their solutions.

AVAILABILITY EQUIVALENCE INCREASING FACTOR

Availability equivalence increasing factor, in short AEIF, referred as $\sigma = \sigma_{S,B}^{D}$, D = H, W, C for hot, warm and cold, respectively, is defined as the factor σ by which the failure rate of a set S components should be reduced in order to get equality of the availability of another better design which can be obtained from the original system by assuming hot, warm and cold duplications of a set B components. That is, $\sigma = \sigma_{S,B}^{D}$, for D = H, W, C, is the solution of the following equations in σ.

$$A_{S,\sigma} = A_{B}^{D}, \quad D = H, W, C. \tag{22}$$

In what follows, we give the non-linear equations needed to be solved to get the three possible AEIF's.

1) Hot availability equivalence increasing factor (HAEIF): Substituting Equations (9) and (11) into Equation (22), $\sigma = \sigma_{S,B}^{H}$, is the solution of the following non-linear equation

$$\prod_{i=1}^{n}\left[1-\prod_{j \in S_i}\left(\frac{\sigma}{\sigma+\eta_{ij}}\right)\prod_{j \in S_i}\left(\frac{1}{1+\eta_{ij}}\right)\right] = \prod_{i=1}^{n}\left[1-\prod_{j \in B_i}\left[1-\left(\frac{\eta_{ij}}{1+\eta_{ij}}\right)^{2}\right]\prod_{j \in \bar{B}_i}\left(\frac{1}{1+\eta_{ij}}\right)\right] \tag{23}$$

2) Warm availability equivalence increasing factor (WAEIF): Substituting Equations (9) and (14) into Equation (22), $\sigma = \sigma_{S,B}^{W}$ is the solution of the following equation in σ

$$\prod_{i=1}^{n}\left[1-\prod_{j \in S_i}\left(\frac{\sigma}{\sigma+\eta_{ij}}\right)\prod_{j \in S_i}\left(\frac{1}{1+\eta_{ij}}\right)\right] = \prod_{i=1}^{n}\left[1-\prod_{j \in B_i}\left(\frac{1+\eta_{ij}+\xi_{ij}}{1+\eta_{ij}+\xi_{ij}+\frac{1}{2}\eta_{ij}^{2}+\frac{1}{2}\eta_{ij}\xi_{ij}}\right)\prod_{j \in \bar{B}_i}\left(\frac{1}{1+\eta_{ij}}\right)\right] \tag{24}$$

3) Cold availability equivalence increasing factor (CAEIF): Substituting Equations (9) and (17) into Equation (22), $\sigma = \sigma^C_{S,B}$ is the solution of the following equation in σ,

$$\prod_{i=1}^{n}\left[1-\prod_{j\in S_i}\left(\frac{\sigma}{\sigma+\eta_{ij}}\right)\prod_{j\in \bar{S_i}}\left(\frac{1}{1+\eta_{ij}}\right)\right] = \prod_{i=1}^{n}\left[1-\prod_{j\in B_i}\left(\frac{1+\eta_{ij}}{1+\eta_{ij}+\frac{1}{2}\eta_{ij}^2}\right)\prod_{j\in \bar{B_i}}\left(\frac{1}{1+\eta_{ij}}\right)\right]. \tag{25}$$

The above Equations (23)-(25) have no closed-form solutions in σ, so a numerical technique method to get the value of σ.

NUMERICAL RESULTS

To explain how one can utilize the previously obtained theoretical results, we introduce a numerical example. In such example, we calculate the two different availability equivalence factors of a general repairable parallel- series with n subsystems. Each subsystem consists of m_i, i=1,2,...,n, non-identical components, under the following assumptions:

1. The parallel-series system has two subsystems, n = 2;
2. The subsystems have the components, $m_1 = 1$, $m_2 = 2$ then N = m_1 + $m_2 = 3$;
3. The values of the system parameters γ_{ij}, μ_{ij}, and v_{ij} $(i = 1, 2, j = 1,...,m_i)$ are presented in Table 1.

The objective is to improve the repairable parallel-series system by improving the performance of some components instead of increasing the number of these components.

We give the values of availability of the original system and of the design obtained using the duplication methods for the example considered in this section.

Table 2 shows the availability of the original and improved system obtained from the original system by applying hot, warm and cold duplications using all possible set B components, where $B = B_1 \cup B_2$ and ϕ is the empty set.

From the results shown in Table 2, one can easily see that:

1. $A_S < A_B^W < A_B^H < A_B^C$, for all possible set B components when $\lambda < v$;
2. $A_S < A_B^H < A_B^W < A_B^C$, for all possible set B components when $\lambda > v$;

3) Improving the only one component in subsystem 1, according to the duplication method, provides a better design than that can be achieved by improving one component from the subsystem 2, according to the same method;

4) Duplicating two components, one from each subsystem, produces a better design than that can be obtained by duplicating the two components in subsystem 2, according to the same method; and 5) Cold duplicating all components in the system provides the best design, in the sense of having the highest availability.

Table 1: Set values of the system parameters

i	j	$\lambda < v$			$\lambda > v$		
		λ	v	μ	λ	v	μ
1	1	0.10	0.12	1.1	0.12	0.1	1.1
2	1	0.11	0.13	1.2	0.13	0.11	1.2
	2	0.12	0.14	1.3	0.14	0.12	1.3

Table 2: The availability of the improved system, A_B^D, D = H, W, C

$|B|$	$B = B_1 \cup B_2$	$\lambda < v$				$\lambda > v$			
		A_S	A_B^H	A_B^W	A_B^C	A_S	A_B^H	A_B^W	A_B^C
1	$B_1 = \{1\}; B_2 = \phi$	0.98655	0.99888	0.99879	0.99939	0.98176	0.99821	0.99833	0.99901
	$B_1 = \phi; B_2 = \{1\}$		0.99242	0.99238	0.99267		0.98959	0.98964	0.98997
	$B_1 = \phi; B_2 = \{2\}$		0.99246	0.99242	0.99271		0.98955	0.98960	0.98992
2	$B_1 = \{1\}; B_2 = \{1\}$		0.99937	0.99931	0.99967		0.99898	0.99905	0.99946
	$B_1 = \{1\}; B_2 = \{2\}$		0.99936	0.99932	0.99967		0.99897	0.99905	0.99945
	$B_1 = \phi; B_2 = \{1,2\}$		0.99882	0.99874	0.99936		0.99814	0.99825	0.99897
3	$B_1 = \{1\}; B_2 = \{1,2\}$		0.99990	0.99989	0.99997		0.99982	0.99984	0.99994

We used Mathematica Program System to calculate all possible availability equivalence factors of the studied system. Table 3 and Table 4 give the hot, warm and cold (D = H, W, C) availability equivalence reducing factors, $\rho = \rho_{R,B}^{D}$, and the hot, warm and cold availability equivalence increasing factors, $\sigma = \sigma_{S,B}^{W}$, respectively, for all possible sets R, S and B.

From the results presented in Table 3, Table 4, we can immediately conclude that:

1) Hot duplication of the only one component in subsystem 1, $B_1 = \{1\}$ and $B_2 = \phi$ increases the system availability from $A_S = 0.98655$ to $A_B^H = 0.99888$ $B = B_1 \cup B_2$, see Table 2. The improved system with $A_B^H = 0.99888$ can be achieved by performing one of the following:

a) Reducing the failure rate(s) of (see Table 3): i) the only component in subsystem 1, $R = R_1 \cup R_2$, where $R_1 = \{1\}$ and $R_2 = \phi$, by the HAERF $\rho_{R,B}^{H} = 0.07692$, ii) the only component in subsystem 1 and the first component in subsystem 2, $R_1 = \{1\}$, $R_2 = \{1\}$, by the HAERF $\rho_{R,B}^{H} = 0.13082$, iii) the only component in subsystem 1 and the second component in subsystem 2, $R_1 = \{1\}$, $R_2 = \{2\}$, by the HAERF $\rho_{R,B}^{H} = 0.13139$, iv) the two components in subsystem 2, $R_1 = \phi$ and $R_2 = \{1,2\}$, by the HAERF $\rho_{R,B}^{H} = 0.07384$, v) all the three components, $R_1 = \{1\}$, $R_2 = \{1,2\}$, by the HAERF $\rho_{R,B}^{H} = 0.26678$.

b) Increasing the repair rate(s) of (see Table 4): i) the only component in subsystem 1, $S = S_1 \cup S_2$, where $S_1 = \{1\}$ and $S_2 = \phi$, by the HAEIF $\sigma_{S,B}^{H} = 13.0000$, ii) the only component in subsystem 1 and first component in subsystem 2, $S_1 = \{1\}$, $S_2 = \{1\}$, by the HAEIF $\sigma_{S,B}^{H} = 7.6442$, iii) the only component in subsystem 1 and second component in subsys-

Table 3: The AERF ($\rho_{R,B}^{D}$, D = H, W, C) for different R, B, when $\lambda < v$

$|R|$	$R = R_1 \cup R_2$	$|B| = 1$		
		$B_1 = \{1\}, B_2 = \phi$	$B_1 = \phi, B_2 = \{1\}$	$B_1 = \phi, B_2 = \{2\}$
$\rho_{R,B}^{H}$				
1	$R_1 = \{1\}; R_2 = \phi$	0.07692	0.54214	0.53933
	$R_1 = \phi; R_2 = \{1\}$	NA	0.07746	0.07202
	$R_1 = \phi; R_2 = \{2\}$	NA	0.08333	0.07792
2	$R_1 = \{1\}; R_2 = \{1\}$	0.13082	0.64953	0.64706
	$R_1 = \{1\}; R_2 = \{2\}$	0.13139	0.65015	0.64767
	$R_1 = \phi; R_2 = \{1,2\}$	0.07384	0.53094	0.52811
3	$R_1 = \{1\}; R_2 = \{1,2\}$	0.26678	0.72971	0.72775
$\rho_{R,B}^{W}$				
1	$R_1 = \{1\}; R_2 = \phi$	0.08333	0.54518	0.54214
	$R_1 = \phi; R_2 = \{1\}$	NA	0.08333	0.07746
	$R_1 = \phi; R_2 = \{2\}$	NA	0.08916	0.08333
2	$R_1 = \{1\}; R_2 = \{1\}$	0.14058	0.65219	0.64953
	$R_1 = \{1\}; R_2 = \{2\}$	0.14118	0.65280	0.65015
	$R_1 = \phi; R_2 = \{1,2\}$	0.08002	0.53398	0.53094
3	$R_1 = \{1\}; R_2 = \{1,2\}$	0.27793	0.73182	0.72971

Availability Equivalence Factors of a General Repairable Parallel

$|B| = 2$			$|B| = 3$
$B_1 = \{1\}, B_2 = \{1\}$	$B_1 = \{1\}, B_2 = \{2\}$	$B_1 = \phi, B_2 = \{1,2\}$	$B_1 = \{1\}, B_2 = \{1,2\}$
0.04323	0.04301	0.08092	0.00670
NA	NA	NA	NA
NA	NA	NA	NA
0.07696	0.07660	0.13692	0.01264
0.07735	0.07699	0.13751	0.01272
0.04144	0.04123	0.07769	0.00641
0.19877	0.19826	0.27378	0.07731
0.04707	0.04682	0.08677	0.00777
NA	NA	NA	NA
NA	NA	NA	NA
0.08333	0.08292	0.14577	0.01464
0.08375	0.08333	0.14638	0.01473
0.04513	0.04489	0.08333	0.00744
0.20758	0.20702	0.28375	0.08333

Table 4: The AEIF ($\sigma_{S,B}^{D}$, D = H, W, C) for different S, B, when $\lambda < \nu$

|S|	$S = S_1 \cup S_2$	$|B| = 1$		
		$B_1 = \{1\}, B_2 = \phi$	$B_1 = \phi, B_2 = \{1\}$	$B_1 = \phi, B_2 = \{2\}$
$\sigma_{S,B}^{H}$				
1	$S_1 = \{1\}; S_2 = \phi$	13.0000	1.8445	1.8542
	$S_1 = \phi; S_2 = \{1\}$	NA	12.9091	13.8852
	$S_1 = \phi; S_2 = \{2\}$	N	12.0000	12.8333
2	$S_1 = \{1\}; S_2 = \{1\}$	7.6442	1.5396	1.5455
	$S_1 = \{1\}; S_2 = \{2\}$	7.6110	1.5381	1.5440
	$S_1 = \phi; S_2 = \{1, 2\}$	NA	NA	NA
3	$S_1 = \{1\}; S_2 = \{1, 2\}$	3.7485	1.3704	1.3741
$\sigma_{S,B}^{W}$				
1	$S_1 = \{1\}; S_2 = \phi$	12.0000	1.8343	1.8445
	$S_1 = \phi; S_2 = \{1\}$	NA	12.0000	12.9091
	$S_1 = \phi; S_2 = \{2\}$	NA	11.2152	12.0000
2	$S_1 = \{1\}; S_2 = \{1\}$	7.1132	1.5333	1.5396
	$S_1 = \{1\}; S_2 = \{2\}$	7.0831	1.5319	1.5381
	$S_1 = \phi; S_2 = \{1, 2\}$	NA	NA	NA
3	$S_1 = \{1\}; S_2 = \{1, 2\}$	3.5980	1.3665	1.3704

$|B| = 2$			$|B| = 3$
$B_1 = \{1\}, B_2 = \phi$	$B_1 = \phi, B_2 = \{1\}$	$B_1 = \phi, B_2 = \{2\}$	$B_1 = \{1\}, B_2 = \phi$
23.1343	23.2500	12.3580	149.2960
NA	NA	NA	NA
N	N	N	N
12.9940	13.0549	7.3034	79.1162
12.9283	12.9888	7.2722	78.6321
NA	NA	NA	NA
5.0310	5.0438	3.6525	12.9354
21.2464	21.3602	11.5245	128.6470
NA	NA	NA	NA
NA	NA	NA	NA
12.0000	12.0600	6.8604	68.3015
11.9404	12.0000	6.8318	67.8861
NA	NA	NA	NA
4.8174	4.8305	3.5242	12.0000

$|S|$	$S = S_1 \cup S_2$	$|B| = 1$		
		$B_1 = \{1\}, B_2 = \phi$	$B_1 = \phi, B_2 = \{1\}$	$B_1 = \phi, B_2 = \{2\}$
$\sigma_{S,B}^C$				
1	$S_1 = \{1\}; S_2 = \phi$	24.0000	1.9093	1.9205
	$S_1 = \phi; S_2 = \{1\}$	NA	23.8182	27.6840
	$S_1 = \phi; S_2 = \{2\}$	NA	20.7991	23.6667
2	$S_1 = \{1\}; S_2 = \{1\}$	13.4495	1.5791	1.5859
	$S_1 = \{1\}; S_2 = \{2\}$	13.3809	1.5775	1.5842
	$S_1 = \phi; S_2 = \{1,2\}$	NA	NA	NA
3	$S_1 = \{1\}; S_2 = \{1,2\}$	5.1260	1.3951	1.3993

tem 2, $S_1 = \{1\}$, $S_2 = \{2\}$, by the HAEIF $\sigma_{S,B}^H = 7.6110$, iv) all the three components, $S_1 = \{1\}$, $S_2 = \{1,2\}$, by the HAEIF $\sigma_{S,B}^H = 3.7485$.

2) Warm duplication of the only component in subsystem 1, $B_1 = \{1\}$ and $B_2 = \phi$, increases the system availability from $A_S = 0.98655$ to $A_B^W = 0.99879$, $B = B_1 \cup B_2$ see Table 2. The improved system with $A_B^W = 0.99879$, can be achieved by performing one of the following:

a) Reducing the failure rate(s) of (see Table 3): i) the only component in subsystem 1, $R = R_1 \cup R_2$ where $R_1 = \{1\}$ and $R_2 = \phi$, by the WAERF $\rho_{R,B}^W = 0.08333$, ii) the only component in subsystem 1 and the first component of subsystem 2, $R_1 = \{1\}$, $R_2 = \{1\}$, by the WAERF $\rho_{R,B}^W = 0.14058$,

$|B| = 2$			$|B| = 3$
$B_1 = \{1\}, B_2 = \phi$	$B_1 = \phi, B_2 = \{1\}$	$B_1 = \phi, B_2 = \{2\}$	$B_1 = \{1\}, B_2 = \phi$
44.0802	44.3267	22.7607	504.5480
NA	NA	NA	NA
NA	NA	NA	NA
23.9930	24.1223	12.7973	265.1530
23.8585	23.9870	12.7328	263.4860
NA	NA	NA	NA
6.9816	7.0014	4.9894	23.8707

iii) the only component in subsystem 1 and the second component of subsystem 2, $R_1 = \{1\}$, $R_2 = \{1\}$, by the WAERF $\rho_{R,B}^W = 0.14118$, iv) the two components in subsystem 2, $R_1 = \phi$, $R_2 = \{1,2\}$, by the WAERF $\rho_{R,B}^W = 0.08002$, v) all three components, $R_1 = \{1\}$, $R_2 = \{1,2\}$, by the WAERF $\rho_{R,B}^W = 0.27793$.

b) Increasing the repair rate(s) of (see Table 4): i) the only component in subsystem 1, $S = S_1 \cup S_2$ where $S_1 = \{1\}$ and $S_2 = \phi$, by the WAEIF $\sigma_{S,B}^W = 12.0000$, ii) the only component in subsystem 1 and first component of subsystem 2, $S_1 = \{1\}$, $S_2 = \{1\}$, by the WAEIF $\sigma_{S,B}^W = 7.1132$, iii) the only component in subsystem 1 and second component of subsystem

2, $S_1 = \{1\}$, $S_2 = \{2\}$, by the WAEIF $\sigma_{S,B}^W = 7.0831$, iv) all three compo-nents, $S_1 = \{1\}$, $S_2 = \{1,2\}$, by the WAEIF $\sigma_{S,B}^W = 3.5980$.

3) Cold duplication of the only component in subsystem 1, $B_1 = \{1\}$ and $B_2 = \phi$, increases the system availability from $A_S = 0.98655$ to $A_B^C = 0.99939$, see Table 2. The improved system with $A_B^C = 0.99939$, can be achieved by performing one of the following:

a) Reducing the failure rate(s) of (see Table 3):

i. the only component in subsystem 1, $R = R_1 \cup R_2$ where $R_1 = \{1\}$ and $R_2 = \phi$ by the CAERF $\rho_{R,B}^C = 0.04167$

ii. the only component in subsystem 1 and first component of subsys-tem 2, $R_1 = \{1\}$, $R_2 = \{1\}$, by the CAERF $\rho_{R,B}^C = 0.07435$,

iii. the only component in subsystem 1 and second component of sub-system 2, $R_1 = \{1\}$, $R_2 = \{2\}$, by the CAERF $\rho_{R,B}^C = 0.07473$, iv) the two components in subsystem 2, $R_1 = \phi$, $R_2 = \{1,2\}$, by the CAERF $\rho_{R,B}$ $= 0.03994$, v) all three components, $R_1 = \{1\}$, $R_2 = \{1,2\}$, by the CAERF $\rho_{R,B}^C = 0.19508$.

b) Increasing the repair rate(s) of (see Table 4): i) the only component in subsystem 1, $S = S_1 \cup S_2$ where $S_1 = \{1\}$ and $S_2 = \phi$ by the CAEIF $\sigma_{S,B}^C = 24.0000$, ii) the only component in subsystem 1 and first component of subsystem 2, $S_1 = \{1\}$, $S_2 = \{1\}$, by the CAEIF $\sigma_{S,B}^C = 13.4495$, iii) the only component in subsystem 1 and second component of subsystem 2, $S_1 = \{1\}$, $S_2 = \{2\}$, by the CAEIF $\sigma_{S,B}^C = 13,3809$, iv) all three compo-nents, $S_1 = \{1\}$, $S_2 = \{1,2\}$, by the CAEIF $\sigma_{S,B}^C = 5.2160$.

4) In the same manner, we can illustrate the rest of results shown in Table 3 and Table 4.

5) The notation NA, means that there is no possible equivalence between the two improved systems that can be achieved by reducing (increasing) the failure (repair) rates of the set R(S) of system components and that can be achieved by duplicating elements of set B of system components.

CONCLUSIONS

This paper discusses the availability equivalence factors of a general repairable parallel-series system with independent but non-identical components. The system studied here generalizes several well-known systems such as a repairable parallel-series system with independent and identical components; repairable series and repairable parallel systems with independent and non-identical or identical components. We derived two types of the availability equivalence factors of the system. We presented a numerical example to illustrate how the theoretical results derived in the paper can be applied.

Indeed there are several possible extensions of this work. As an example, the case of a general repairable parallel-series system with non-constant failure rates can be studied.

REFERENCES

1. Sarhan, A.M. (2002) Reliability Equivalence with a Basic Series/Parallel System. Applied Mathematics and Computation, 132, 115-133. http://dx.doi.org/10.1016/S0096-3003(01)00181-3

2. Sarhan, A.M. (2009) Reliability Equivalence Factors of a General Series-Parallel System. Reliability Engineering and System Safety, 94, 229-236.http://dx.doi.org/10.1016/j.ress.2008.02.021

3. Sarhan, A.M., Al-Ruzaiza, A.S., Alwasel, I.A. and El-Gohary, A.I. (2004) Reliability Equivalence of a Series-Parallel System. Applied Mathematics and Computation, 154, 257-277. http://dx.doi.org/10.1016/S0096-3003(03)00709-4

4. R?de, L. (1993) Reliability Equivalence. Microelectronics Reliability, 33, 323-325.http://dx.doi.org/10.1016/0026-2714(93)90020-Y

5. Meng, F.C. (1993) On Selecting Components for Redundancy Incoherent Systems. Reliability Engineering and System Safety, 41, 121-126. http://dx.doi.org/10.1016/0951-8320(93)90025-T

6. Mustafa, A. (2009) Reliability Equivalence Factor of n-Components Series System with Non-Constant Failure Rates. International Journal of Reliability and Applications, 10, 43-58.

7. Hu, L., Yue, D. and Zhao, D. (2012) Availability Equivalence Analysis of a Repairable Series-Parallel System. Mathematical Problems in Engineering, 2012, Article ID: 957537.http://dx.doi.org/10.1155/2012/957537

8. Sarhan, A.M. and Mustafa, A. (2013) Availability Equivalence Factors of a General Repairable Series-Parallel System. International Journal of Reliability and Applications, 14, 11-26.

9. Lewis, E.E. (1996) Introduction to Reliability Engineering. 2nd Edition, Wiley, New York.

10. Leemis, L.M. (1996) Reliability Probabilistic Models and Statistical Methods. Prentice-Hall, Englewood Cliffs.

11. Kumar, S., Chattopadhyayb, G. and Kumar, U. (2007) Reliability Improvement through Alternative Designs—A Case Study. Reliability Engineering and System Safety, 92, 983-991. http://dx.doi.org/10.1016/j.ress.2006.05.008

12. Billinton, R. and Wang, P. (1999) Deregulated Power System Planning Using a Reliability Network Equivalent Technique. IEE Proceedings—Generation, Transmission and Distribution, 146, 25-30. http://dx.doi.org/10.1049/ip-gtd:19990046

13. Sarhan, A.M., Tadj, L., Al-Khedhairi, A. and Mustafa, A. (2008) Equivalence Factors of a Parallel-Series System. Applied Sciences, 10, 219-230.

14. Xia, Y. and Zhang, G. (2007) Reliability Equivalence Factors in Gamma Distribution. Applied Mathematics and Computation, 187, 567-573.http://dx.doi.org/10.1016/j.amc.2006.07.016

15. Mustafa, A. and El-Bassoiuny, A.H. (2009) Reliability Equivalence of Some Systems with Mixture Linear Increasing Failure Rates. Pakistan Journal of Statistics, 25, 149-163.

16. Mustafa, A. and El-Faheem, A.A. (2012) Reliability Equivalence Factors of a General Parallel System with Mixture Lifetimes. Applied Mathematical Sciences, 6, 3769-3784.

17. Mustafa, A., Sarhan, A.M. and Al-Ruzaiza, A.S. (2007) Reliability Equivalence of a Parallel-Series System. Pakistan Journal of Statistics, 23, 241-254.

18. Wang, Z.H. (1992) Reliability Engineering Theory and Practice. Quality Control Society of Republic of China, Taipei, China.

19. Juang, Y.S., Lin, S.S. and Kao, H.P. (2008) A Knowledge Management System for Series-Parallel Availability Optimization and Design. Expert Systems with Applications, 34, 181-193. http://dx.doi.org/10.1016/j.eswa.2006.08.023

20. Kolowrocki, K. (1994) Limit Reliability Functions of Some Series-Parallel and Parallel-Series Systems. Applied Mathematics and Computation, 62, 129-151. http://dx.doi.org/10.1016/0096-3003(94)90078-7

21. Liu, Y. and Zheng, H. (2010) Study on Reliability of Warm Standby's Repairable System with n Identity Units and k Repair Facilities. Journal of Wenzhou University, 31, 24-29.

22. Gu, J. and Wei, Y. (2006) Reliability Quantities of a n-Unit Cold Standby Repairable System with Two Repair Facility. Journal of Gansu Lianhe University, 20, 17-20.

CITATION

Mustafa, A. and Sarhan, A. (2014) Availability Equivalence Factors of a General Repairable Parallel-Series System.Applied Mathematics, 5, 1713-1723. doi: 10.4236/am.2014.511164.

The Summation of One Class of Infinite Series

Jonathan D. Weiss
JSA Photonics LLC, Corrales, New Mexico, USA

ABSTRACT

This paper presents closed-form expressions for the series, $\sum(-1)^{n+1}\coth(nx)/n$, where the sum is from $n = 1$ to $n = \infty$. These expressions were obtained by recasting the series in a different form, followed by the use of certain relationships involving the elliptical nome. Among the values of x for which these expressions can be obtained are of the form: $2^l\pi\lambda$ and $2^l\pi/\lambda$, where l is an integer between $-\infty$ and ∞. The values of λ include 1, $\sqrt{2}, \sqrt{3},$, and 3. Examples of closed-form expressions obtained in this manner are first presented for $x = \pi/4$, $\pi/2$, π, and 2π. Additional examples are then presented for $x = \pi/\left(2\sqrt{3}\right)$, $\pi/\left(2\sqrt{2}\right)$, $\pi/\sqrt{2}$, and $\pi\sqrt{3}/2$. This undertaking was prompted by the author's work on an electrostatics boundary-value problem related to the van der Pauw measurement technique of electrical resistivity. The presence of this series for $x = \pi$ in the solution of that problem and its absence from any compendium of infinite series that he consulted led to this work.

INTRODUCTION

The series S(x), where

$$S(x) = \sum_{n-1}^{\infty} (-1)^{n+1} \coth(nx)/n \tag{1}$$

does not appear in any of several compendiums of infinite series or products [1] -[4] . The author first encountered it for $x = \pi$, as part of a solution to an electrostatics boundary-value problem, and he deduced a simple analytic expression for $S(\pi)$ based on the known solution to this problem in a limiting case. This is opposed to a "first-principles" derivation. The problem involves the determination of the electrical resistivity of a uniform material in the shape of a thin square of side "a" and thickness c, as $c \rightarrow 0$. This situation arises in the semiconductor industry in the electrical characterization of blank wafers of materials, such as silicon, within which numerous circuits are fabricated.

The particular measurement configuration considered here is shown in Figure 1.

Depicted in this figure there is a square sample of side "a" and a square contact array of semi-diameter w, rotated by 45° with respect to the sample and displaced from the center of the sample by the vector s. This displacement can be in any direction and magnitude as long as the array remains completely within the sample (including on the boundary). Current I_o is forced to enter contact 1 and leave contact 2, and the resulting potential difference between contacts 4 and 3, $\Delta\varphi = \varphi_4 - \varphi_3$, is determined from a voltage measurement. Mathematically, the contacts are considered point-like, which in practice means that their diameter $\bullet w$ (the same is also true for c). The practical purpose of solving the boundary-value is to determine the voltage and thus the stability of the results with respect to the displacement and w [5] . In [5] , these results were compared to those associated with a non-rotated array. When s = 0 and w/a = 1/2, we obtain the situation shown in Figure 2, where all four contacts are on the boundary in a highly symmetric arrangement.

This is a special case of the more general one considered by van der Pauw [6] in a seminal paper in which neither the sample nor the contact array is required to have any simple shape, only that the sample be thin, have no holes in it, and that all contacts are point-like and on its periphery. He used the method of conformal mapping, not the electrostatics method of [5] . For the particular configuration in Figure 2, his results reduce to:

$$\rho = \left(\frac{\Delta\varphi}{I_o}\right)\left(\frac{c\pi}{\ln(2)}\right)$$

(2)

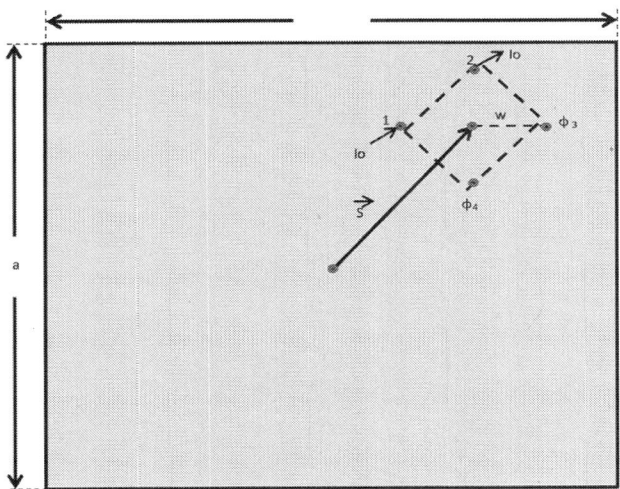

Figure 1: General configuration of the electrostatics boundary-value problem that led to the series in question.

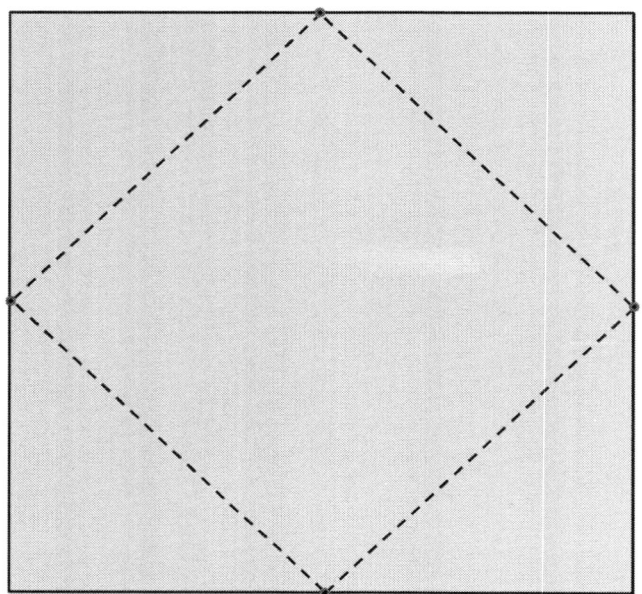

Figure 2: The limiting case of the configuration in Figure 1, or the symmetric van der Pauw arrangement, satisfied by Equation (5). The current and voltage points are not shown.

In this equation, ρ is the electrical resistivity. When w/a < 1/2 and s ≠ 0, Equation (2) must be modified, such that:

$$\rho = \left(\frac{\Delta\varphi}{I_o}\right)\left(\frac{c\pi}{F\left(\dfrac{w}{a}, s\right)}\right)$$

(3)

Clearly, F(1/2, 0) = ln(2). The function F(w/a, s) was calculated in [5] using a standard separation-of-va- riables technique applied to Laplace's equation for the potential everywhere within the sample, along with the appropriate surface boundary conditions on its gradient. The function $F(w/a, 0)$ is as follows:

$$F\left(\frac{w}{a},0\right) = \sum_{n=1}^{\infty} g_1 \left[g_2 - g_3 - g_4 + g_5\right]:$$

(4)

$$g_1 = \frac{1}{2n\sinh n\pi}$$

$$g_2 = \left[\cosh\frac{n\pi w}{a} + \cosh n\pi\left(1-\frac{w}{a}\right)\right]\left[\cos\frac{n\pi w}{a} + \cos n\pi\left(1-\frac{w}{a}\right)\right]$$

$$g_3 = \left[1+\cosh n\pi\right]\left[(-1)^n + \cos\frac{n\pi 2w}{a}\right]$$

$$g_4 = \left[1+\cosh n\pi\left(1-\frac{2w}{a}\right)\right]\left[1+(-1)^n\right]$$

$$g_5 = \left[\cosh\frac{n\pi w}{a} + \cosh n\pi\left(1-\frac{w}{a}\right)\right]\left[\frac{n\pi w}{a} + \cos n\pi\left(1+\frac{w}{a}\right)\right]$$

Allowing $w/a = 1/2$, we obtain, after considering even and odd n separately:

$$F\left(\frac{1}{2},0\right) = \sum_{n=1}^{\infty}(-1)^{n+1}\frac{\coth(n\pi)}{n} - \sum_{n=1}^{\infty}1/(n\sinh(2n\pi))$$

(5)

Using the relationship [4] :

$$\sum_{n=1}^{\infty} 1/\left(n\sinh\left(2n\pi\right)\right) = \frac{\pi}{6} - \left(\frac{3}{4}\right)\ln\left(2\right)$$

(6)

and F(1/2, 0) = ln(2), we obtain:

$$\sum_{n=1}^{\infty}\left(-1\right)^{n+1}\frac{\coth\left(n\pi\right)}{n} = \frac{\pi}{6} + \left(\frac{1}{4}\right)\ln\left(2\right)$$

(7)

It is clear that the sum in Equation (7) should be very close to ln(2) because coth(nπ) → 1 so quickly with n. In fact, the right-hand side of Equation (7) is 0.6968×××, while ln(2) = 0.6931×××, resulting in a percentage difference of about 0.5%. When the left-hand side was summed directly, the two sides agreed to the limit of precision of the software used, or 15 significant figures.

ALTERNATE DERIVATION FOR SEVERAL VALUES OF X

The above derivation is based on a requirement of a physics problem and applies to only one value of x, x = π. The following derivation, which applies to an array of values of x, proceeds by recasting the series in a different form and applying certain existing relationships to evaluate the resulting series.

Using the definition of the hyperbolic cotangent, we rewrite S(x) as:

$$S\left(x\right) = \sum_{n=1}^{\infty}\left(-1\right)^{n+1}\frac{\left(1+e^{-2nx}\right)}{n\left(1-e^{-2nx}\right)}$$

(8)

This series can be expressed in a more useful form for our current purpose. We first note that for $|y| < 11/\left(1-y\right) = \sum y^{k}$, where the sum is from $k = 0$ to $k = \infty$, and that $\ln(1+z) = \Sigma(-1)^{m+1} z^m / m$, where $-1 < z \leq 1$ and the sum is from $m = 1$ to $m = \infty$. In this case, $y = e^{-2nx}$ and $z = 1$. Consequently, after a few simple manipulations, we can express $S(x)$ as:

$$S(x) = 2\sum_{k=1}^{\infty} \ln\left(1 + e^{-2kx}\right) + \ln(2)$$

(9)

For certain values of x, the series can now be evaluated in a straight forward manner, using existing relationships.

EVALUATION OF THE SERIES

These relationships are stated in [4] ("Series of Logarithms" and "Specific Values" of "Series Expansion of Inverse Elliptical Nome Q"). Obtained from "Series of Logarithms" is the most fundamental of these relationships for our purpose:

$$\sum_{k=1}^{\infty} \ln\left(1 + q^k\right) = \left(\frac{1}{24}\right) \ln\left[m \Big/ \left(16q(1-m)^2\right)\right].$$

(10)

In this expression, q is the elliptical nome [7] and m is the inverse elliptical nome ($0 < m < 1$). In the second section of [4], certain values of q are stated along with their associated m. All of these expressions for q are of the form: $e^{-\pi\lambda}$, where λ is real and >0 (this must be true of λ, based on the definition of q in [7]). Thus, if we are to use these results, $x = \pi\lambda/2$, although it is not clear why or if x must be some multiple of π to obtain an analytic expression for $S(x)$. Since Equation (10) is fundamental to the evaluation of $S(x)$, some background information concerning it is called for. First, it is equivalent to:

$$\prod_{k=1}^{\infty} \left(1 + q^k\right) = \left[m \Big/ 16q(1-m)^2\right]^{1/24}$$

(11)

Second, in [4] under "Inverse Elliptic Nome Q m[q] and K [m[q]] Expressed through Infinite Products" are the following relationships:

$$m = 1 - \prod_{k=1}^{\infty}\left[\left(1-q^{2k-1}\right)\big/\left(1+q^{2k-1}\right)\right]^{8} = 16q\prod_{k=1}^{\infty}\left[\left(1+q^{2k}\right)\big/\left(1+q^{2k-1}\right)\right]^{8}$$
(12)

Using this equation to form the right-hand side of Equation (11) easily leads to its left-hand side, after a few simple algebraic manipulations. The equality between the first and second terms in Equation (12) is discussed in [8] , while that between the first and third is discussed in [9] .

Also in the second section of [4] are three relationships that allow us to expand the number of values of x ad infinitum. These relationships are as follows:

$$m\left(q^{2}\right) = \left[\frac{1-\sqrt{1-m(q)}}{1+\sqrt{1-m(q)}}\right]^{2}$$
(13)

or its inverse,

$$m(q) = \frac{4\sqrt{m\left(q^{2}\right)}}{\left(1+\sqrt{m\left(q^{2}\right)}\right)^{2}}$$
(14)

and

$$1 - m\left(e^{-\pi\lambda}\right) = m\left(e^{-\frac{\pi}{\lambda}}\right)$$
(15)

The first of these three equations allows one to generate the inverse nome of q^2, given the inverse nome of q, and that of q^4 from that of q^2, etc. From Equation (14), the inverse nome of $q^{1/2}$, $q^{1/4}$, etc. can be generated starting with that of q. Thus, starting with a particular value of X, say X_0, one can generate an entire array of X, say X_n, to which Equation (10) can be applied, such that:

$$X_n = 2^n X_0; \quad n = -\infty, \cdots, -2, -1, 0, 1, 2, \cdots, \infty$$

(16)

By using Equation (15), which follows directly from the definition of elliptical nome in [7] (ch. 17), one can generate additional values of X_0 through the inverse of λ (i.e., $X_0 = \pi\lambda/2 \rightarrow X_0 = \pi/2\lambda$). This has been done for a few cases in [4]. It also follows directly from Equation (15) that m = 1/2 for $\lambda = 1$.

EXAMPLES OF CLOSED-FORM SOLUTIONS

A few examples of the use of these equations to generate closed-form expressions for S(x) will now be presented. They will be for $\lambda = 1/2$, 1, 2, and 4, which corresponds to x = $\pi/4$, $\pi/2$, π, and 2π, respectively. Since the value of S(x) for x = π motivated this entire undertaking, it is included here. Starting with $\lambda = 1$, for which m = 1/2, one can use Equation (14) to generate the value of m for $\lambda = 1/2$ and Equation (16) to generate m for $\lambda = 2$ and 4. The value of m for $\lambda = 1/2$ can also be generated from the value of m for $\lambda = 2$, using Equation (15). All of these results are actually presented in [4]. The other starting values of λ in [4] that have been used to generate the remaining results are: $\sqrt{2}$ and $\sqrt{3}$.

$x = \pi/4 \, ; \, q = e^{-\pi/2} \, ; \, m = 12\sqrt{2} - 16$

$$S\left(\frac{\pi}{4}\right) = \left(\frac{\pi}{24}\right) + \left(\frac{5}{4}\right)\ln 2 + \left(\frac{1}{4}\right)\ln\left(1 + \frac{3\sqrt{2}}{4}\right) = 1.178\cdots$$

(17)

$x = \pi/2 \, ; \, q = e^{-\pi} \, ; \, m = 1/2$

$$S\left(\frac{\pi}{2}\right) = \left(\frac{\pi}{12}\right) + \left(\frac{3}{4}\right)\ln 2 = 0.781\cdots$$

$$(18)$$

$$\underline{x = \pi;\, q = e^{-2\pi;}\,;\, m = 17 - 12\sqrt{2}}$$

$$S(\pi) = \left(\frac{\pi}{6}\right) + \left(\frac{1}{4}\right)\ln(2) = 0.696\cdots$$

$$(19)$$

$$\underline{x = 2\pi;\, q = e^{-4\pi;}\,;\, m = \left[\left(2^{1/4} - 1\right)\big/\left(2^{1/4} + 1\right)\right]^{4}}$$

$$S(2\pi) = \left(\frac{\pi}{3}\right) - \left(\frac{1}{4}\right)\ln(2) - \left(\frac{1}{4}\right)\ln\left(1 + \frac{3\sqrt{2}}{4}\right) = 0.69315\cdots$$

$$(20)$$

$$\underline{x = \pi\big/2\sqrt{3}\,;\, q = e^{-\pi/\sqrt{3};}\,;\, m = \left(2 + \sqrt{3}\right)\big/4}$$

$$S\left(\frac{\pi}{2\sqrt{3}}\right) = \left(\frac{\pi}{12\sqrt{3}}\right) + \left(\frac{5}{6}\right)\ln 2 + \left(\frac{1}{12}\right)\ln\left[\frac{2 + \sqrt{3}}{7 - 4\sqrt{3}}\right] = 1.058\cdots$$

$$(21)$$

$$\underline{x = \pi\big/2\sqrt{2}\,;\, q = e^{-\pi/\sqrt{2};}\,;\, m = \left(2\sqrt{2} - 2\right)}$$

$$S\left(\frac{\pi}{2\sqrt{2}}\right) = \left(\frac{\pi}{12\sqrt{2}}\right) + \left(\frac{5}{6}\right)\ln 2 + \left(\frac{1}{12}\right)\ln\left[\frac{\sqrt{2} - 1}{17 - 12\sqrt{2}}\right] = 0.925\cdots$$

$$(22)$$

$$\underline{x = \pi\big/\sqrt{2}\,;\, q = e^{-\pi\sqrt{2};}\,;\, m = 2 - 2\sqrt{2}}$$

$$S\left(\frac{\pi}{\sqrt{2}}\right) = \left(\frac{\pi\sqrt{2}}{12}\right) + \left(\frac{1}{2}\right)\ln 2 = 0.716\cdots$$

$$\tag{23}$$

$$x = \pi\sqrt{3}/2; \; q = e^{-\pi\sqrt{3}}; \; m = \left(2 - \sqrt{3}\right)/4$$

$$S\left(\frac{\pi\sqrt{3}}{2}\right) = \left(\frac{\pi\sqrt{3}}{12}\right) + \left(\frac{5}{6}\right)\ln 2 + \left(\frac{1}{12}\right)\ln\left[\frac{2 - \sqrt{3}}{7 + 4\sqrt{3}}\right] = 0.701\cdots$$

$$\tag{24}$$

All of these results (and others not shown) have been confirmed by an agreement with a direct summation of each series. Figure 3 is a graphical representation S(x) vs. x/π. We note that, as expected, S(x) rapidly approaches ln(2) with increasing x. At the other extreme, S(x) → ∞ as (1/x) for small x. The inset in Figure 3 demonstrates this behavior on a log-log plot.

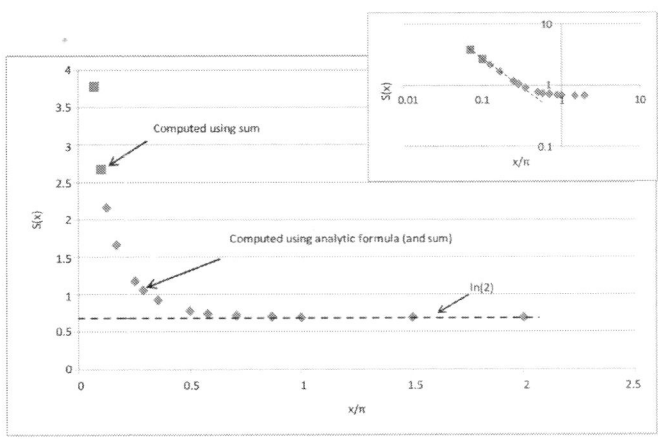

Figure 3: Graphical representation of the results of this study. The inset is a log-log scale in which the dotted line is an eyeball fit.

CONCLUSION

Simple analytic formulas have been produced for eight values of x for the series $\sum (-1)^{n+1} \coth(nx)/n$, where the sum is from $n = 1$ to $n = \infty$, although the sum could also be over the negative integers. These values are a subset of a much larger set of x for this possible. This larger set obviously encompasses the corresponding negative values of x. The formulas were produced by recasting the series into a form such that known relationships involving the elliptical nome could be applied to their derivation. The author first encountered the series for $x = \pi$ in an electrostatics boundary-value problem. The sum for this value of x was the result of the application of the series to a limiting case of that problem. This value of x was also one of the eight just mentioned. It is quite possible that this series, with the appropriate values of x, is at least part of the solution of other physical problems.

ACKNOWLEDGEMENTS

The author is grateful to J. Mercer of Albuquerque, New Mexico for reviewing the manuscript and making valuable suggestions for its improvement. This work was self-funded.

REFERENCES

1. Hansen, E.H. (1970) A Table of Series and Products. Prentice Hall, Englewood Cliffs.
2. Prudnikov, A.P., Brychkov, Yu.A. and Marichev, O.I. (1986) Integrals and Series. Elementary Functions Vol. 1. Gordon and Breach, New York.
3. Gradshteyn, I.S. and Ryzhik, I.M. (1965) Table of Integrals, Series, and Products. Academic Press, New York.
4. Dieckmann, A. (2000) Collection of Infinite Products and Series. http://pi.physik.uni-bonn.de/~dieckman/InfProd/InfProd.html
5. Weiss, J.D. (2014) A Comparison of Two Van-der-Pauw Measurement Configurations. Materials Science in Semiconductor Processing, Submitted.

6. Van der Pauw, L.J. (1958) A Method of Measuring the Resistivity and Hall Coefficient on Lamellae of Arbitrary Shape. Philips Technical Review, 20, 220-224.
7. Abramowitz, A. and Stegun, I.A. (1972) Handbook of Mathematical Functions. Dover Publications, New York.
8. Whittaker, E.T. and Watson, G.N. (1996) A Course of Modern Analysis. 4th Edition, Cambridge University Press, Cambridge. http://dx.doi.org/10.1017/CBO9780511608759
9. Erdélyi, A., Ed. (1955) Higher Transcendental Functions (Vol. III). Mc Graw Hill, New York.

CITATION

Weiss, J. (2014) The Summation of One Class of Infinite Series. Applied Mathematics, 5, 2815-2822. Doi: 10.4236/am.2014.517269.

A Complete and Simple Solution to A Discrete-Time Finite-Capacity BMAP/D/C Queue

Nam K. Kim[1], Mohan L. Chaudhry[2], Bong K. Yoon[3], and Kilhwan Kim[4*]

[1]Department of Industrial Engineering, Chonnam National University, Gwangju, South Korea
[2]Department of Mathematics and Computer Science, Royal Military College of Canada, Kingston, Canada
[3]Defense Management College, Korea National Defense University, Seoul, South Korea
[4]Department of Management Engineering, Sangmyung University, Cheonan, South Korea

8

ABSTRACT

We consider a discrete-time multi-server finite-capacity queueing system with correlated batch arrivals and deterministic service times (of single slot), which has a variety of potential applications in slotted digital telecommunication systems and other related areas. For this queueing system, we present, based on Markov chain analysis, not only the steady-state distributions but also the transient distributions of the system length and of the system waiting time in a simple and unified manner. From these distributions, important performance measures of practical interest can be easily obtained. Numerical examples concerning the superposition of certain video traffics are presented at the end.

INTRODUCTION

The discrete-time multi-server queue with deterministic service times has gained importance in view of a number of potential practical applications to slotted digital telecommunication systems and other related areas (Bruneel and Wuyts [1]). It has been observed that arrival streams to these systems, in particular, tend to be correlated (see, e.g., Wittevrongel and Bruneel [2,3]). To model this correlated nature, a versatile point process called the discrete-time batch Markovian arrival process (D-BMAP) is introduced by Blondia and Casals [4] and widely used for analytical studies. This rich class of arrival processes contains a number of well-known arrival processes such as the Bernoulli process with independent identically distributed batch arrivals, the Markov modulated Bernoulli process, and a superposition of D-BMAP themselves (for more details, see Blondia and Casals [4]).

In this paper, we consider the D-BMAP/D/c/N queue, in which customers arrive according to D-BMAP and are served by one of c servers. The system capacity is $N \geq c$ so that no more than N customers can be accommodated in the system at the same time. Customers who arrive to find the system full are assumed to depart the system immediately on arrivals. Specifically, we assume a partial acceptance model (Takagi [5], p. 367) such that customers of an arriving batch are accepted until they fill all the available capacity, and the remaining customers, if any, are lost. Every accepted customer requires one slot for service that is assumed to start and end at slot boundaries.

For this queueing model, we present, based on an elementary Markov chain analysis, both steady-state and transient solutions to the system length (i.e., the number of customers in system) as well as to the system waiting time (i.e., the number of slots a customer spends in system) in a simple and unified manner. For similar discrete-time multi-server queueing models with infinite capacity, there have been some contributions by other authors: For a similar queue with infinite capacity (and correlated batch arrivals that belong to a subclass of B-DMAP arrivals), Sohraby and Zhang [6] analyze in transform domain the transient behavior of the system length and present an efficient numerical inversion method to calculate a few performance measures of interest. (They briefly discuss the finite capacity case as well.) For the D-BMAP/D/c queue, Alfa [7] assumes constant service times of multiple slots and presents an efficient algorithm making clever use of the structural property of the system to obtain the steady-state distributions of the system length and the system waiting time. Gao et al. [8] (Gao et al. [9]) assume constant service times of multiple slots (geometric service times) with a two-state Markovian arrival process to present a steady-state analysis of the system length and the system waiting time.

In this paper, we assume the finite-capacity model for the following three practical reasons. First and foremost, as demonstrated in this paper, the finite-capacity model of this paper is much simpler to analyze than its corresponding infinite-capacity counterpart (see Remark 1 below; also, see Sohraby and Zhang [6] for sophisticated analysis of the infinite-capacity model). Second, queueing models with finite capacity can serve as excellent approximations (by taking the system capacity N sufficiently large) for their corresponding infinite-capacity counterparts (see Remark 2 at the end of this paper). Third, all the queueing systems in reality have finite capacity.

We organize the paper as follows: In Section 2, we first present the steady-state system-length distribution based on the elementary Markov chain analysis. From this, important performance measures of practical interest are obtained, including the steady-state system waiting time distribution. In addition, the corresponding transient solutions are presented in a simple and unified manner. In Section 3, we present a set of numerical results with various system capacities and a few different numbers of servers. For this, we use D-BMAP arrivals that characterize the superposition of certain video traffics. We end the paper with a remark on the finite-capacity model.

ANALYSIS

In discrete-time queueing models, the time axis is divided into fixed-length intervals, called slots. It is assumed that customer arrivals and departures take place only at slot boundaries; thus, nothing is assumed to happen in the middle of a slot.

In the D-BMAP/D/c/N queue, customers arrive according to a D-BMAP with representation $\{D_k, k \geq 0\}$, where D_k is an $m \times m$ matrix with elements

$(D_k)_{ij} \leq 1, j \leq m$. This arrival process is governed by an m-state (or m-phase) underlying Markov chain (UMC). Specifically, let us suppose that the UMC is in some phase i in a certain slot. Then, with probabilities $(D_k)_{ij}$, there are $k \geq 0$ arrivals during the slot with the phase of the UMC being j in the next slot. (See Blondia and Casals [4] for more details.) Note that the number of arrivals per slot (including those who are lost) is given by

$$\lambda = \pi \sum_{k=1}^{\infty} kD_k e$$

(1)

where π is the stationary probability vector of the UMC with the transition probability matrix (TPM) $\sum_{k=0}^{\infty} D_k$ and e is a column vector of 1's (note that π is obtained by solving simultaneously the equations $\pi = \pi$ D and $\pi e = 1$ for π).

Now, we consider the discrete-time bivariate process $\{(N_k, S_k); 0 \leq N_k \leq N, 1 \leq S_k \leq m, k \geq 1\}$, where N_k and S_k denote, respectively, the system length and the phase of the UMC just after the beginning of the kth slot. Then we consider the number of customers that arrive during the kth slot (denoted by A_k) and the number of customers that depart at the end of the same slot (which is given by $\min(N_k, c)$); as a result, we have N_{k+1} in terms of N_k as follows:

$$N_{k+1} = \min\left(N_k + A_k, N\right) - \min\left(N_k, c\right), k \geq 1 .$$

(2)

Note that A_k is dependent only on S_k; thus, we obtain the discrete-time Markov chain having the following TPM: (please see the formula below)

Where $D_{\geq l} = D_l + D_{l+1} + \ldots, l \geq 0$.

Let $p^{(k)} = (p_0^{(k)}, p_1^{(k)}, \cdots, p_N^{(k)})$ denote the state probability vector of the bivariate process just after the beginning of the kth slot, where

$$p_n^{(k)} = \left(p_{n,1}^{(k)}, \cdots, p_{n,m}^{(k)}\right)$$

And $p_{n,i}^{(k)} = \Pr\{(N_k, S_k) = (n,i)\}$. Then it is immediate to have $P^{(k+1)} = P^{(k)} \cdot T, k \geq 1$.

Steady-State Analysis

1) The System-Length Distribution: Let

$p = (p_0, p_1, \cdots, p_N) = \lim_{k \to \infty} p^{(k)}$. Then the steady-state system-length distribution is obtained by solving simultaneously the equations $p = pT$ and $pe = 1$ for p. This can be readily carried out by using mathematical software packages such as MATLAB, Mathematica, or even otherwise (see numerical examples in Section 3). Now, important performance measures of practical interest can be obtained from P_n as given below.

$$
T = \begin{array}{c} 0 \\ 1 \\ \vdots \\ c \\ c+1 \\ \vdots \\ N-1 \\ N \end{array}
\left(\begin{array}{ccccccccc}
\boldsymbol{D}_0 & \boldsymbol{D}_1 & \cdots & \boldsymbol{D}_{N-c-1} & \boldsymbol{D}_{N-c} & \cdots & \boldsymbol{D}_{N-1} & \boldsymbol{D}_{\geq N} \\
\boldsymbol{D}_0 & \boldsymbol{D}_1 & \cdots & \boldsymbol{D}_{N-c-1} & \boldsymbol{D}_{N-c} & \cdots & \boldsymbol{D}_{N-1} & 0 \\
\vdots & \vdots & \ddots & \vdots & \vdots & \ddots & \vdots & \vdots \\
\boldsymbol{D}_0 & \boldsymbol{D}_1 & \cdots & \boldsymbol{D}_{N-c-1} & \boldsymbol{D}_{\geq N-c} & 0 & \cdots & 0 \\
0 & \boldsymbol{D}_0 & \cdots & \boldsymbol{D}_{N-c-2} & \boldsymbol{D}_{\geq N-c-1} & 0 & \cdots & 0 \\
\vdots & \vdots & \vdots & \ddots \, \boldsymbol{D}_{N-c-2} & \vdots & \vdots & \ddots & \vdots \\
0 & 0 & \cdots & \boldsymbol{D}_0 & \boldsymbol{D}_{\geq 1} & 0 & \cdots & 0 \\
0 & 0 & \cdots & 0 & \boldsymbol{D}_{\geq 0} & 0 & \cdots & 0
\end{array} \right)
\tag{3}
$$

2) The Effective Arrival Rate and the Loss Probability: The number of arrivals (including those who are lost) per slot is given by

$$
\lambda = \sum_{n=0}^{N} P_n \sum_{k=1}^{\infty} k D_k e .
\tag{4}
$$

(Equation (4) reduces to (1) due to $\sum_{n=0}^{N} P_n = \pi$). Under the assumption of partial acceptance, the number of accepted arrivals per slot (i.e., the effective arrival rate), on the other hand, is given by

$$\lambda_e = \sum_{n=0}^{N-1} p_n \left(\sum_{k=1}^{N-n} kD_k + \sum_{k=N-n+1}^{\infty} (N-n)D_k \right) e$$

$$= \sum_{n=0}^{N-1} p_n \left(\sum_{k=1}^{N-n} D_{\geq k} \right) e.$$

(5)

From (4) and (5), it is immediate to have the loss probability (i.e., the probability that a customer is lost) as follows:

$$P_{\text{loss}} = 1 - \frac{\lambda_e}{\lambda}.$$

(6)

3) Moments of the System Length: Among others, the first moments of the numbers of customers in system and in service just after a slot boundary, denoted by L and L_s respectively, are given by

$$L = \sum_{n=1}^{N} np_n e$$

(7)

and

$$L_S = \sum_{n=1}^{c} np_n e + \sum_{n=c+1}^{N} cp_n e.$$

(8)

Along the same lines, higher moments such as the variances of the numbers of customers in system and in service can be easily obtained.

Note that L_s also represents the number of departures (by service completions) per slot, i.e., the departure rate, which, of course, should match with the effective arrival rate; that is, $L_s = \lambda_e$. Thus, the loss probability can be alternatively obtained from $P_{loss} = 1 - L_s / \lambda$.

4) The System Waiting-Time Distribution of an Accepted Customer: Let W denote the system waiting time of an accepted customer, i.e., the number of slots an accepted customer spends in system (we do not count, as a part of the system waiting time, the slot in which she arrives). The probability that an accepted customer spends at most w slots in system can be interpreted as the long-run fraction of such customers out of all accepted customers. That is,

$$P(W \le w) = E(A(w))/\lambda_e$$

(9)

where $A(w)$ denotes the number of accepted customers per slot who are to spend at most w slots in system. Note that because the service times are single slot, one can foresee whether the system waiting time of an accepted customer will exceed w or not. Let N_∞ and S_∞ denote, respectively, the system length and the phase of the UMC just after the beginning of a slot at steady state. Then, conditioning on N_∞ and S_∞, we have $E(A(w))$ as follows;

$$E(A(w)) = \sum_{n=0}^{N} \sum_{i=1}^{m} E(A(w)|N_\infty = n, S_\infty = i) \cdot p_{n,i} ,$$

,(10)

where

$$E\left(A(w)\middle|N_\infty = n, S_\infty = i\right)$$

$$= \begin{cases} \sum_{j=1}^{m}\left\{ \sum_{k=1}^{\min(cw,N-n)} k\left(D_k\right)_{i,j} \right. \\ \left. + \sum_{k=\min(cw,N-n)+1}^{\infty} \min\left(cw, N-n\right)\left(D_k\right)_{i,j} \right\} \\ \quad 0 \le n \le c, \\ \sum_{j=1}^{m}\left\{ \sum_{k=1}^{\min(c(w+1)-n,N-n)} k\left(D_k\right)_{i,j} \right. \\ \left. + \sum_{k=\min(c(w+1)-n,N-n)+1}^{\infty} \min\left(c(w+1)-n, N-n\right)\left(D_k\right)_{i,j} \right\} \\ \quad c+1 \le n \le N-1. \end{cases}$$

After simplifications (i.e., following the same procedure as used in getting the last term of (5)), we have

$$E\left(A(w)\right) = \sum_{n=0}^{c} p_n \sum_{k=1}^{\min(cw,N-n)} D_{\ge k}e \\ + \sum_{n=c+1}^{N-1} p_n \sum_{k=1}^{\min(c(w+1)-n,N-n)} D_{\ge k}e.$$

(11)

Substituting (11) into (9), we have the steady-state system waiting-time distribution of an accepted customer. From this, one can get performance measures of interest, such as the mean $(E(W))$ of the system waiting time and its tail probabilities. Also, one can get L alternatively by Little's formula, $L = \lambda_e \cdot E(W)$.

Transient Analysis

Note that $p^{(k)}$ is obtained from $P^{(k)}=P^{(1)} \cdot T^{k-1}, k \geq 1$, where $P^{(1)}$, the initial probability vector, is assumed to be given. Putting $P_n^{(k)}$ in place of P_n of (4) through (11) derived for the steady state, one can immediately obtain the corresponding transient results: the expected numbers of total, accepted, and lost arrivals during the k th slot, the moments of the numbers of customers in system as well as in service just after the beginning of the same slot, and the system waiting-time distribution of customers that are accepted during that slot.

Remark 1: Transient analyses of queueing models are, in general, much more demanding than their stationary counterparts, because the former need to take an additional variable (time) into consideration. See, e.g., Sohraby and Zhang [6] for a transient analysis of the queue of a similar kind with infinite capacity. This is not the case for the finite-capacity case, which can be analyzed in a remarkably simple and unified manner as presented in this paper.

NUMERICAL EXAMPLES

For numerical work, we use the same D-BMAP arrival as the one given in Example 2 of Blondia and Casals [4]. In this example, they approximate the superposition of 3 video sources by the superposition of 30 independent identical on/off sources. The latter is then characterized by the D-BMAP, where the phase of the UMC corresponds to the number of active sources. (See Blondia and Casals [4] for the representation of this D-BMAP.)

For various system capacities $N\{4,6,8,10,12\}$ with different numbers of servers $c=\{1,2,3\}$, Table 1 gives the steady-state loss probabilities. From this, one can see how fast the loss probability decreases as the system capacity N increases and as the number of servers c increases.

Table 2, with fixed capacity $N=10$ and the numbers of servers $c=\{1,2,3\}$, gives the steady-state probabilities that the system waiting time of an accepted customer exceeds given threshold values $\{1,2,3,4,5\}$. Such tail probabilities are one of the important performance measures of practical interest (particularly, in telecommunication area) that represent the quality of service.

For $p_n^{(1)} = 0, 0 \le n \le N-1$, and $p_N^{(1)} = \pi$ (i.e., the system is initially full of customers) with fixed capacity N=10 and the numbers of servers c={1,2,3}, Figure 1 displays how fast the expected system length of each system just after the beginning of the kth slot (k=1,2,...20) converges to its corresponding steadystate quantity. Other measures of practical interest can be obtained along the same lines.

Remark 2: For each system with the number of ser- vers c={1,2,3}, we observe that the loss probability tends to converge to zero pretty quickly with a moderate increase in the system capacity (see Table 1); in addition, we observe that the tail probability of each waiting-time distribution decays pretty quickly as well (see Table 2). This seems to be mainly due to the extreme regularity of the (constant) service times and the multiple numbers of servers, both of which, individually as well as jointly, absorb burstiness of the arrival process considerably. Consequently, in such cases, one can effectively reduce the loss probability, the tail probability of waiting-time distribution, or both with a slight increase in the system capacity or the number of severs. Besides, in such cases as the loss probabilities are practically zero, a finite- capacity model can serve as an excellent approximation for the corresponding infinite-capacity counterpart. Then one can avoid sophisticated analyses for the infinite-

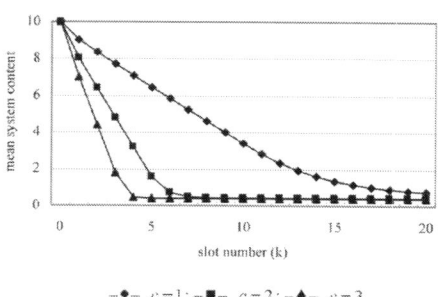

Figure 1: Transient mean system lengths.

Table 1: Steady-state loss probabilities

Number of severs	Sys capacity					
	N.4	N.6	N s8	N-10	N.12	
c =1	7.388627E-03	6.05570E-04	6.98359E-05	1.09435E-05	2.21558E-06	
c = 2	3.682965E-03	6.69900E-05	8.55853E-07	1.01555E-08	1.35632E-10	
c = 3	3.611292E-03	5.86295E-05	5.58459E-07	3.60230E-09	3.01454E-11	

Table 2: Steady-state tail probabilities of the system waiting time

Number of savers	P (W > I)	P (W > 2)	P(W > 3)	P(W > 4)	P (W > 5)
c =1	0.262522	0.062382	0.015792	0.004463	0.001390
c = 2	0.021443	0.000175	1.23E-06	2.64E-10	0
com 3	0.001665	2.98E-07	4.31E-12	0	0

capacity model and get both steady-state and transient solutions in a remarkably simple and unified manner as presented in this paper.

We hope that the elementary Markov-chain based analysis we present in this paper for the finite-capacity D-BMAP/D/c/N queue would turn out to be beneficial to both theoreticians and practitioners who would like simple and straightforward practical solutions to their complex queueing systems.

ACKNOWLEDGMENTS

The first author acknowledges with thanks the support provided by Prof. Sungjune Park and BISOM (Business Information Systems and Operations Management) of University of North Carolina at Charlotte, where he held an Adjunct-Visiting Professorship during his sabbatical year 2012 and where part of this work was done. The second author's research was supported in part by grant 4010 NSERC RG.

REFERENCES

1. A H. Bruneel and I. Wuyts, "Analysis of Discrete-Time Multiserver Queueing Models with Constant Service Times," Operations Research Letters, Vol. 15, No. 5, 1994, pp. 231-236.doi:10.1016/0167-6377(94)90082-5
2. S. Wittevrongel and H. Bruneel, "Exact Calculation of Buffer Contents Variance and Delay Jitter in a Discrete-Time Queue with Correlated Input Traffic," Electronics Letters, Vol. 32, No. 14, 1996, pp. 1258-1259. doi:10.1049/el:19960848
3. S. Wittevrongel and H. Bruneel, "Discrete-Time Queues with Correlated Arrivals and Constant Service Times," Computers & Operations Research, Vol. 26, No. 2, 1999, pp. 93-108. doi:10.1016/S0305-0548(98)00053-7
4. C. Blondia and O. Casals, "Statistical Multiplexing of VBR Sources: A Matrix-Analytic Approach," Performance Evaluation, Vol. 16, No. 1-3, 1992, pp. 5-20. doi:10.1016/0166-5316(92)90064-N
5. H. Takagi, "Queueing Analysis," Vol. 3, Discrete-Time Systems, North-Holland, Amsterdam, 1993.

6. K. Sohraby and J. Zhang, "Spectral Decomposition Approach for Transient Analysis of Multi-Server DiscreteTime Queues," Performance Evaluation, Vol. 21, No. 1-2, 1994, pp. 131-150. doi:10.1016/0166-5316(94)90031-0

7. S. Alfa, "Algorithmic Analysis of the BMAP/D/k System in Discrete Time," Advances in Applied Probability, Vol. 35, No. 4, 2003, pp. 1131-1152. doi:10.1239/aap/1067436338

8. P. Gao, S. Wittevrongel and H. Bruneel, "On the Behavior of Multiserver Buffers with Geometric Service Times and Bursty Input Traffic," IEICE TRANSACTIONS on Communications, Vol. E87-B, No. 12, 2004, pp. 3576- 3583.

9. P. Gao, S. Wittevrongel, J. Walraevens and H. Bruneel, "Analytic Study of Multiserver Buffers with Two-State Markovian Arrivals and Constant Service Times of Multiple Slots," Mathematical Methods of Operations Research, Vol. 67, No. 2, 2008, pp. 269-284.doi:10.1007/s00186-007-0163-z

CITATION

N. Kim, M. Chaudhry, B. Yoon and K. Kim, "A Complete and Simple Solution to a Discrete-Time Finite-Capacity BMAP/D/c Queue," Applied Mathematics, Vol. 3 No. 12A, 2012, pp. 2169-2173. Doi: 10.4236/am.2012.312A297.

A Trend-Based Segmentation Method and the Support Vector Regression for Financial Time Series Forecasting

Jheng-Long Wu and Pei-Chann Chang

Department of Information Management,
Yuan Ze University, Taoyuan 32026, Taiwan

ABSTRACT

This paper presents a novel trend-based segmentation method (TBSM) and the support vector regression (SVR) for financial time series forecasting. The model is named as TBSM-SVR. Over the last decade, SVR has been a popular forecasting model for nonlinear time series problem. The general segmentation method, that is, the piecewise linear representation (PLR), has been applied to locate a set of trading points within a financial time series data. However, owing to the dynamics in stock trading, PLR cannot reflect the trend changes within a specific time period. Therefore, a trend based segmentation method is developed in this research to overcome this issue. The model is tested using various stocks from America stock market with different trend tendencies. The experimental results show that the proposed model can generate more profits than other models. The model is very practical for real-world application, and it can be implemented in a real-time environment.

INTRODUCTION

Support vector machines (SVMs) have outperformed other forecasting models of machine learning or soft computing (SC) tools such as decision tree, neural network (NN), bayes classifier, fuzzy systems (FSs), evolutionary computation (EC), and chaos theory by many researchers from historical nonlinear time series data applications in the last decade [1–5]. In these techniques, many researchers presented different forecasting models in dealing with characteristics such as imprecision, uncertainty, partial truth, and approximation to achieve practicability, robustness, and low solution cost in real applications [6–8]. However, the most important issue in resolving the nonlinear time series problem is error revision. ANNs use the empirical risk minimization principle to minimize the generalization errors but SVRs use the structural risk minimization principle because SVR is able to analyze with small samples and to overcome the local optimal solution problem, which surpasses to ANNs [9–11]. Therefore, the SVRs forecasting model is applied to accomplish the forecasting task in this research. Presently, support vector regression (SVR), which was evolved from support vector machine (SVM) based on the statistical learning theory, is a powerful forecasting and machine learning approach for numerical prediction [12–15]. Also, SVR has high toleration error rate and high accuracy for learning solution knowledge in complex problems [16]. Although SVR can be applied well in time series data, the input vector is a key successful factor. Despite the volatile nature of the stock markets, researchers still can find certain correlations between these factors and stock prices. An investor's primary goal is to make profits. In order to help investors achieve their financial objectives, researchers have studied the relationship between financial markets and price variations over time from [17–20].

In the last few years, several representations of time series data have been proposed; the most often used representation is piecewise linear representation (PLR) [21–23]. It can decompose a time series data into a series of bottom and peak points [24, 25] in financial market. But the traditional PLR does not consider the multiple trending characteristics in time series. Moreover, the price movements of stocks are affected by many factors such as government policies, economic environments, interest rates, and inflation rates. The share prices of most listed companies also move up and down with other changing factors like market capitalization, earnings per share (EPS), price- to -earnings ratio, demand and supply, and market news. Moreover, there are more fractal properties of financial data, such as self-similarity, heavy-tailed distributions, long memory, as well as power laws [26–29]. One of fractal properties is long memory which is a common characteristic in financial data or other fields [30–32]. The daily stock trading is a short-term return so in this paper these fractal properties were not considered in our framework, just focusing on the real stock price's trends.

Therefore, there is a need to develop a new segmentation method which takes the price moving trends into consideration. As a result, this research will consider the multiple trends of stock price's movements in TBSM segmentation approach to capture the embedded knowledge of nonlinear time series. This research intends to improve the SVR forecasting performance using a trend based decomposition method. The TBSM approach has captured the tendency of stock price's movement which can be inputted into SVR in learning the historical knowledge of the time series data. Moreover, a more accurate forecasting result can be achieved when applied in real-time stock trading decision.

The rest of this paper is organized as follows. In Section 2, we describe TBSM segmentation principle. Forecasting model is discussed in Section 3. Section 4 explains modeling for trading decisions including using historical data to make trading decisions by the TBSM approach, selecting highly correlated technical indices by stepwise regression analysis (SRA), forecasting trading signals by SVR, and evaluating trading strategies. Section 5 explains how the TBSM with SVR for stock trading decisions and compares the profits obtained from various forecasting approaches. Finally, conclusions and directions for further research are discussed in Section 6.

A TREND BASED SEGMENTATION METHOD (TBSM)

In the time series database there are many approaches such as Fourier transform, wavelets, and piecewise linear representation which can be applied to find the turning point on time series data. According to the characteristics of sequential data, a piecewise linear representation of the data is more appropriate. A variety of algorithms to obtain a proper linear representation of segment data have been presented. As reported in [33–36], PLR is used to support more tasks and provides an efficient and effective solution. In this paper we intend to enhance the segmentation accuracy based on different trends in stock price's movements. The basic idea of TBSM is to modify the PLR segmentation using the trend tendency in a specific time period. Three different trends such as uptrend, downtrend, and hold trend will be considered when making the segmentation. Detailed procedures of TBSM include the following. (1) PLR is applied to locate the turning points from the time series including up or downtrends. (2) The points around each turning point will be double-checked if the variations of the points are within the threshold. If yes, these points will have the same buy/sell trading in this period. (3) These points are set to be in the same trend. The pseudocode of the TBSM is shown in Algorithm 1.

Define: Threshold // cutting threshold	
	X_Thld // horizontal area
	Y_Thld // vertical area
	X // a time series
	Y // stock price
1:	Procedure TBSM(T)
2:	Let T be represented as X[1, 2,..., n], Y[1, 2,..., n]
3:	n = 0
4:	Draw a line between (X_1, Y_1) and (X_n, Y_n)
5:	Max d = maximum distance of (X_i, Y_i) to the line

6:	If (Max d > Threshold)
7:	Let (X_i, Y_i) be the point with maximum distance
8:	For j = X_1:X_n
9:	If $(\lvert X_j - X_i \rvert < X\_Thld)$ and $(\lvert Y_j - Y_i \rvert < Y\_Thld)$
10:	Then Point[n]=$[X_j,X_j]$, n = n + 1
11:	End If
12:	End For
13:	Select from Point[n]:X_{t1}=Min(X_0),X_{t2}=Max(X_n)
14:	Return: S1 = T $[X_1,X_{t1}]$
15:	S2 = T $[X_{t2},X_n]$
16:	End If

Algorithm 1: A pseudocode for TBSM in time series data.

For example, a time series T=$\{t_1, t_2, \ldots, t_{191}\}$ with 191 data is given to explain the basic idea of the TBSM procedure. As shown in Figure 1(a), several trading points are represented as buy (four red points) or sell (six green points) in this case. According to the TBSM procedure, we can draw a line S_1 form the first point to the last point as shown in Figure 1(b) and find the max distance to line S_1 which is point t_{26}. Then line S_1 is decomposed into two segments including line S_2 from t_1 to t_{26} and line S_3 from t_{26} to t_{19}1. Based on point t_{26}, we can locate point t_{16} to t_{56} which are varied within the threshold. These points are set as hold trend andwith the same state of point t_{26}. Therefore line S_2 and line S_3 will be changed to three different lines including line S_4 from point t_1 to point t_{16}, line S_5 from point t_{16} to point t_{56}, and line S_6 is from point t_{56} to point t_{191} as shown in Figure 1(c). Next step is repeating the same process for the rest of segments as t_{56} to t_{191}. The final results are shown in Figure 1(d) including two hold trend segments (dotted line), one uptrend segment, and two downtrend segments (solid line) in this time series.

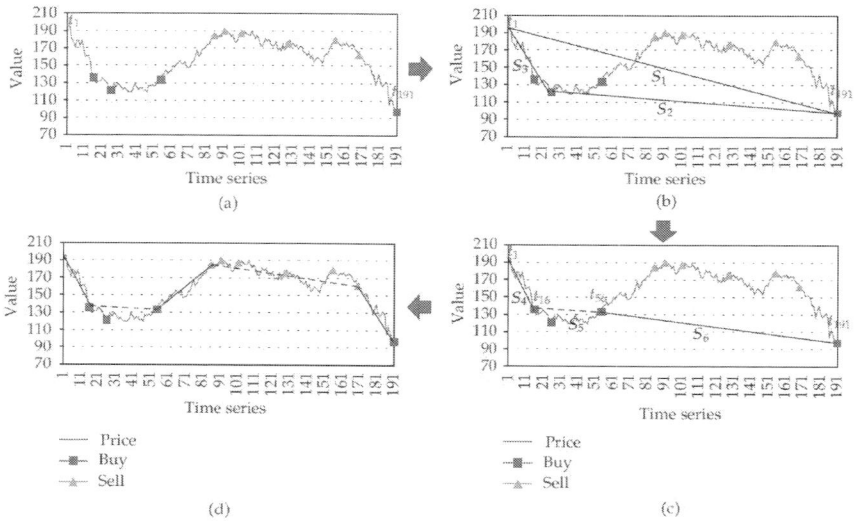

Figure 1: An example for TBSM in time series data.

SUPPORT VECTOR REGRESSIONS (SVRS)

Support vector regression is a modification of machine-learning-theory-based classification called support vector machine. Machine learning techniques have been applied for assigning trading signal. Many studies used support vector machine for determining whether a case contains particular class [37, 38]. But the shortcoming only deal with discrete class labels, whereas trading signal continuum data type because a weight of signal can take a buy or sell power. Grounded in statistical learning theory [1, 2], support vector regression is capable to predict the continuous trading signal while still benefiting from the robustness of SVM. SVM has been successfully employed to solve forecasting problems in many fields, such as financial time series forecasting [39] and emotion computation [40]. For explaining the concept of SVR, we have considered a standard regression problem. Let $S=\{X_i, Y_i\}_{i=1...n}$ be the set of data where Xi is input vector (selected technical index in this research), Y_i (trading signal t_s) is an output vector, and n is the number of data points. In regression analysis, we find a function $f(X_i)$ such that $Y_i = f(X_i)$. This function can be used to find the output value Y of any X. The standard regression function is as follows:

$$q_i = f(x_i) + \delta, \tag{3.1}$$

where δ denotes the random error and q_i denotes the estimated output. There are two types of regression problems, namely, linear and nonlinear. SVR is developed to tackle the nonlinear regression problems because the nonlinear regression problems have high complexity as well as stock market trade. In SVR, at first the input vectors are non-linearly mapped into a high-dimensional feature space (F), where they are linearly correlated with the respective output values.

SVR uses the following linear estimation function:

$$f(x) = (\omega \cdot \phi(x)) + b, \tag{3.2}$$

where ω denotes the weight vector, b denotes a constant, $\phi(x)$ denotes the mapping function in the feature space, $(\omega \cdot \phi(x))$ and denotes the dot product in the feature space F SVR transfers the nonlinear regression problem of the lower dimension input space (x) into a linear regression problem of a high-dimension feature space. In other words, the optimization problem involving a nonlinear regression is converted into finding the flattest function in the feature space instead of input space.

Various cost functions like Laplacian, Huber's Gaussian, and ε insensitive can be used in the formulation of SVR. The cost function should be suitable for the problem and should not be very complicated because a complicated cost function could lead to difficult optimization problems. Thus, we have used robust ε-sensitive cost function which is shown below:

$$L_\varepsilon(f(x), q) = \begin{cases} |f(x) - q| - \varepsilon, & \text{if } |f(x) - q| \geq 0 \\ 0, & \text{otherwise,} \end{cases} \tag{3.3}$$

where ε denotes a precision parameter which represents the radius of the tube located around the regression function.f(x).

The {+ε,-ε}region is called ε insensitive zone. ε is determined by the user. If the actual output value lies in this region, the forecasting error is considered to be zero.

The weight vector, ω, and constant, b, in (3.2) are calculated by minimizing regularized risk function which is shown in (3.4):

$$R(C) = \frac{C}{n} \sum_{i=1}^{n} L_\varepsilon \left(f(x_i), q_i \right) + \frac{1}{2} |\omega|^2,$$

(3.4)

where $L_\varepsilon (f(x_i), q_i)$ denotes the -insensitive loss function, $|\omega^2| / 2$ denotes the regularization term, and denotes the regularization constant. ω decides the complexity and approximate accuracy of the regression model. Value of C is selected by the user to ensure appropriate value of w and low empirical risk.

The two positive slack variables ξ_i and ξ_i^* are used to replace the -insensitive loss function of (3.3). ξ_i is defined as the distance between the q_i and higher boundary of the ε insensitive zone, and ξ_i^* is defined as the distance between the q_i and lower boundary of the -insensitive zone. Equation (3.4) is transformed into (3.5) by using the slack variables:

$$\text{Minimize} : R_{\text{reg}}(f) = \frac{1}{2}|\omega|^2 + C \sum_{i=1}^{n} (\xi_i + \xi_i^*)$$

(3.5)

$$\text{Subject to} \begin{cases} q_i - (\omega \cdot \phi(x_i)) - b \leq \varepsilon + \xi_i \\ (\omega \cdot \phi(x_i)) + b - q_i \leq \varepsilon + \xi_i^* \\ \xi_i, \xi_i^* \geq 0, \quad \text{for } i = 1, \ldots, n. \end{cases}$$

(3.6)

Lagrange function method is used to find the solution which minimizes the regression risk of (3.4) with the cost function in (3.3) which results in the following quadratic programming problem (QP):

$$\text{Minimize}: \frac{1}{2}\sum_{i=1}^{N}\sum_{j=1}^{N}(\alpha_i - \alpha_i^*)(\alpha_j - \alpha_j^*)(\phi(x_i) \cdot \phi(x_j))$$

$$+ \sum_{i=1}^{N}\left(\varepsilon_i^{up} - y_i\right)\alpha_i + \sum_{i=1}^{N}\left(\varepsilon_i^{down} - y_i\right)\alpha_i^*, \tag{3.7}$$

$$\text{Subject to}: \sum_{i=1}^{N}(\alpha_i - \alpha_i^*) = 0, \quad \text{where } \alpha_i, \alpha_i^* \in [0, C], \tag{3.8}$$

where α_i and α_i^* denote Lagrange multipliers. ε_i^{up} and ε_i^{down} represent the i th up- and downmargin, respectively. The value of ε_i^{up} and ε_i^{down} is equal to ε. The QP problem of (3.7) is solved under the constraints of (3.8). After solving the QP problem, we obtained Lagrange multiplier from (3.9), and (3.2) is transformed into the following equation (3.10):

$$w = \sum_{i=1}^{N}(\alpha_i - \alpha_i^*) \cdot \phi(x_i), \tag{3.9}$$

$$f(x) = (\alpha_i - \alpha_i^*)(\phi(x_i) \cdot \phi(x)) + b. \tag{3.10}$$

The Karush-Kuhn-Tucker (KKT) conditions are used to find the value of b KKT conditions state that at the optimal solution, the product between the Lagrange multipliers and the constraints is equal to zero. The value of b can be calculated as follows:

$$b = \begin{cases} y_i - (w \cdot \phi(x_i)) - \varepsilon_i^{up}, & \text{for } \alpha_i \in (0, C), \\ y_i - (w \cdot \phi(x_i)) + \varepsilon_i^{down}, & \text{for } \alpha_i^* \in (0, C). \end{cases}$$

$$(3.11)$$

Using the trick of the kernel function, (3.10) can be written as (3.12):

$$f(x) = \sum_{i=1}^{n} (\alpha_i - \alpha_i^*) K(x, x_i) + b,$$

$$(3.12)$$

where $K(x, x_i) = (\phi(x) \cdot \phi(x_i))$ denotes the kernel function which is symmetric and satisfies the Mercer's condition. SVR was able to predict the nonlinear relationship between technical indices and trading signal ts better than other soft computing (SC) techniques.

APPLICATION IN FINANCIAL TIME SERIES DATA

This paper proposes a forecasting framework using a TBSM combined with SVR model which is called TBSM-SVR trading model for stock trading. The framework of TBSM-SVR trading model has five stages: the first is generating nonlinear trading segments by TBSM approach from historical stock price; the second is trading signal transformation from trading segments; the third is feature selection from technical indices by SRA approach; the fourth is learning the trading forecasting model by SVRs approach. The framework of TBSM-SVR model is shown in Figure 2. The five stages of TBSM-SVR model are explained as follows.

Find Turning Points Based on Multiple Trend by TBSM

According to TBSM procedure to find turning point based on trend of stock price, we selected a time series of historical stock price in a period to segment into several segments based on three trends including uptrend, downtrend, and hold trend. For example, a time series is given to segment trend segments from the date 2008/1/2 to 2008/12/30. Figure 3 shows the segmentation result by our proposed TBSM approach. The blue line is original historical stock price. The dashed lines are up/down a trend which if the segment trend goes up is belonging to uptrend and if the segment trend goes down is belonging to downtrend. The dot line is belonging to hold trend. In our experiment, each stock price can split to multiple trend segments for trading signal transformation.

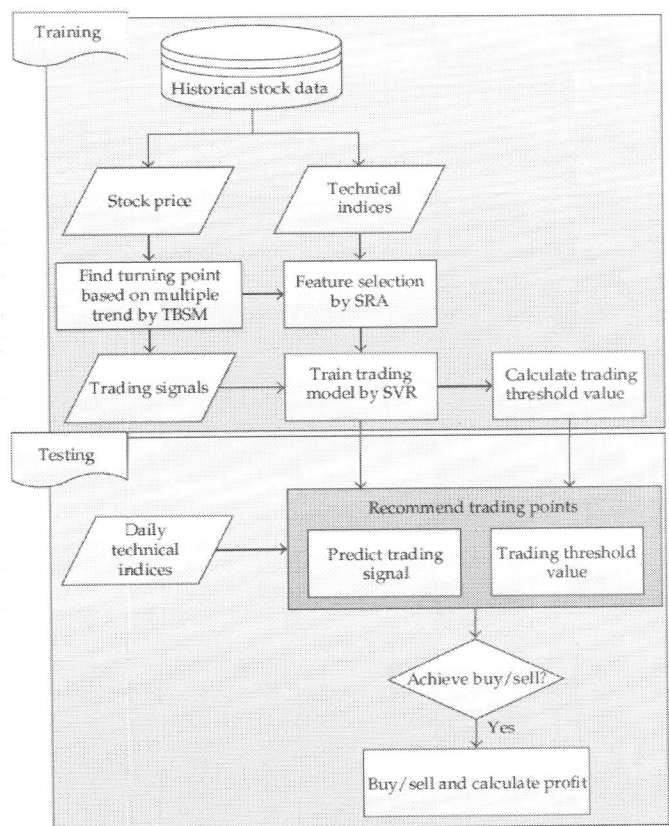

Figure 2: The framework of TBSM-SVR model for stock trading.

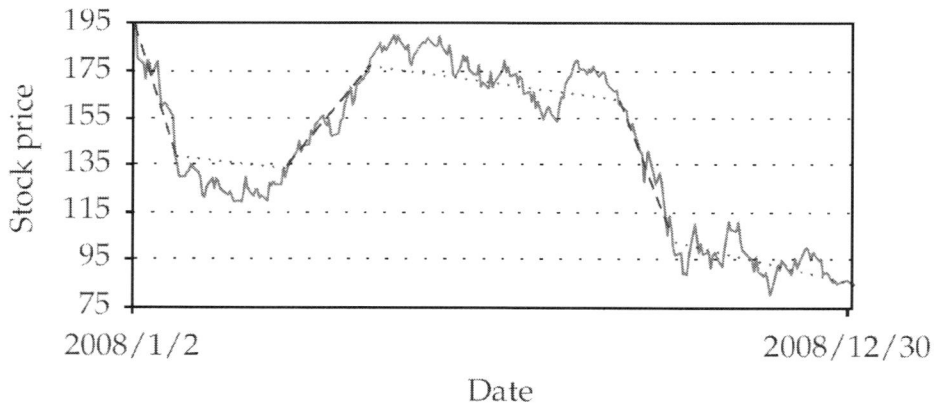

Figure 3: An example of segmentation result by TBSM.

Trading Signal Transformation

In this stage, the aim is calculating the trading signal for a nonlinear time series of segmentation result which are a lot of segments based on trends. We suppose a segment S_k is uptrend; then we assume the real value into the vector S'_k like to $S_k = [0, 0.1,..., 1]$; if S_k is hold trend but locates in buy point, then the vector like to $S'_k = [0.5, 0, 0.5]$; if S_k is hold trend but locates in sell point; then the vector like to $S'_k = [0.5, 1, 0.5]$; if S_k is downtrend, then the vector S'_k like to $[1, 0.9,..., 0]$. Finally we combine these S'_k to a full time series of trading signal ts. If the segment belongs to uptrend or downtrend, then the formula equation (4.1) is used to calculate trading signal value:

$$S'_{k,i} = \begin{cases} \dfrac{i}{L} & \text{if } S_k \text{ is uptrend segment,} \\[2mm] \dfrac{(L-i)}{L} & \text{if } S_k \text{ is downtrend segment,} \end{cases} \tag{4.1}$$

where L denotes the length of segment S_k, whereas segment belonging to hold trend is using (4.2) to calculation:

$$S'_{k,i} = \begin{cases} 1 & \text{if } i\text{th is higherpoint in time series,} \\ 0 & \text{if } i\text{th is lower point in time series,} \\ 0.5 & \text{otherwise.} \end{cases}$$

$$(4.2)$$

For example, the S_1, and S_3 are hold trend; the S_1 is down-trend; the S_4 is up-trend. The result of trading signal t_s is shown in Figure 4. The red dotted line is the hold trend which is a special signal for increasing reflects on the original turning points, so the hold trend is not a horizontal line. The purple dotted line is downtrend signal, and the orange dotted line is uptrend signal. For example, in the time series T the T_1 to T_5 and T_{10} to T_{14} are hold trend signal representation, T_6 to T_9 is downtrend signal representation, and finally T_{15} to T_{18} is uptrend signal representation. Finally the trading signal t_s which is like to $t_s = \{S_1, S_2, S_3, S_4\} = \{\langle 0.5,0.5,1,0.5,0.5 \rangle, \langle 1,0.66,0.333,0 \rangle, \langle 0.5,0.5,0,0.5,0.5 \rangle, \langle 0,0.33,0.66,1 \rangle\}$. For the detail process see the pseudocode in Algorithm 2.

Input: length, oldTs // input data length and vector.
Output: newTs // a new time series vector of trading signal.
Method:
1:   Start = oldTs 1
2:   End = oldTs[length]
3:   If Start = = −1 and End = = 1
4:       newTs 1 = 0
5:       For i = 1: length−1
6:         newTs[i+1] = 1/(lenghth−1)*i
7:       End For
8:   Else If Start = = 1 and End = = −1
9:       newTs[length] = 0
10:      For i = 1 : length−1
11:        newTs[i+1] = 1/(lenghth−1)*(length−i)
12:      End For

13:	Else
14:	For i = 2 : length−1
15:	newTs[i] = 0.5
16:	End For
17:	End If

Algorithm 2: A pseudocode for trend segments by TBSM in time series.

Feature Selection for Technical Indices by SRA

In this paper, we have considered 28 variables (technical indices) as listed in Table 1. These variables are correlated with variations in stock prices to some degree. The quantity of correlation varies for different variables. Rather than using all the 28 variables, we select the variables with a greater correlation than a user-defined threshold. The variable selection is done by stepwise regression analysis. We apply the SRA approach to determine which technical indices affecting the stock price. This is accomplished by selecting the variables repeatedly.

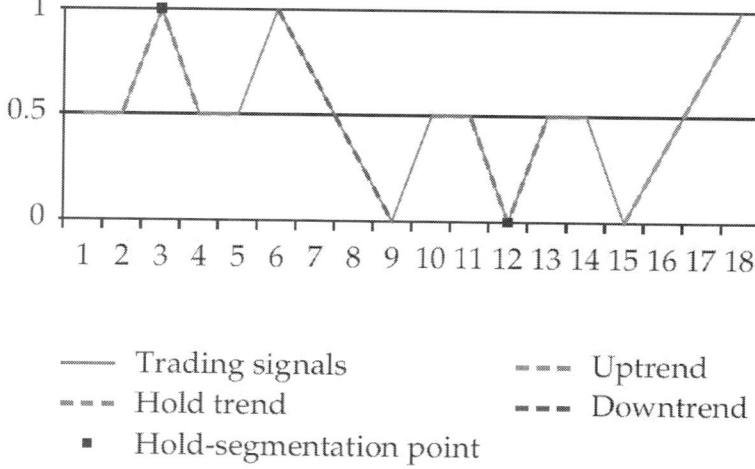

Figure 4: A sample of trading signal.

In the feature selection part input factors will be further selected using stepwise regression analysis (SRA). The SRA has been applied to determine the set of independent variables which is most closely affecting the dependent variable. The SRA is step by step to select factor into regression model which if factor has the significance level, then it is selected. We can follow (4.4) to calculate the F value of SRA:

$$SSR = \sum \left(\hat{Y} - \overline{Y} \right)^2,$$

$$SSE = \sum \left(\hat{Y}_i - Y_i \right)^2,$$

(4.3)

$$F_j^* = \frac{MSR(x_j \mid x_i)}{MSE(x_j \mid x_i)} = \frac{SSR(x_j \mid x_i)}{SSE/(n-2)\ (x_j \mid x_i)} \quad i \in I,$$

(4.4)

where SSR denotes a regression sum of square. SSE denotes residual sum of squares. x is the value of technical index. y is the value of stock price. n is the total number of training data. $\hat{Y}$ is the forecasting value of regression. $\overline{Y}$ is the average stock price of training data. After the feature selection by SRA, we can provide a set of features to form an input vector for the next step to learning the forecasting model.

The steps of the SRA approach are described as follows.

Step 1. Find the correlation coefficient r for each technical index $\upsilon_1, \upsilon_2, ..., \upsilon_n$ with the stock price y in a stock. These correlation coefficients are stored in a matrix called correlation matrix.

Step 2. The technical index with largest R^2 value is selected from the correlation matrix. Let the technical index be υ_i. Derive a regression model between the stock price and technical index, that is $\hat{y} = f(\upsilon_i)$.

Step 3. Calculate the partial F value of other technical indices. Compare the R^2 value of the remaining technical indices and select the technical index with the highest correlation coefficient. Let the technical index be υ_j. Derive another regression model, that is $\hat{y} = f(\upsilon_i, \upsilon_j)$.

Table 1: Technical indices used as input variables

Technical	Technical index	Explanation
Moving average (Ma)	5 MA, 6 MA, 10 MA, 20 MA	Moving averages are used to emphasize the direction of a trend and smooth out price and volume fluctuations that can confuse interpretation.
Bias (BIAS)	5 BIAS, 10 BIAS	The difference between the closing value and moving average line, which uses the stock price nature of returning back to average price to analyze the stock market.
Relative strength index (RSI)	6 RSI, 12 RSI	RSI compares the magnitude of recent gains to recent losses in an attempt to determine overbought and oversold conditions of an asset.
Nine days stochastic line (K, D)	9 K, 9 D	The stochastic line K and line D are used to determine the signals of overpurchasing, overselling, or deviation.
Moving average convergence and divergence (MACD)	9 MACD	MACD shows the difference between a fast and slow exponential moving average (EMA) of closing prices. Fast means a short-period average, and slow means a long period one.
Williams %R (pronounced "percent R")	12 W%R	Williams %R is usually plotted using negative values. For the purpose of analysis and discussion, simply ignore the negative symbols. It is best to wait for the security's price to change direction before placing your trades.

Technical	Technical index	Explanation
Williams %R (pronounced "percent R")	12 W%R	Williams %R is usually plotted using negative values. For the purpose of analysis and discussion, simply ignore the negative symbols. It is best to wait for the security's price to change direction before placing your trades.
Transaction volume (TV)	5 TV, 10 TV, 15 TV	Transaction volume is a basic yet very important element of market timing strategy. Volume provides clues as to the intensity of a given price move.
Differences of technical index (Δ)	$\Delta 5\,MA$, $\Delta 6\,MA$, $\Delta 10\,MA$, $\Delta 5\,BIAS$, $\Delta 10\,BIAS$, $\Delta 6\,RSI$, $\Delta 12\,RSI$, $\Delta 12\,W\%R$, $\Delta 9K$, $\Delta 9D$, $\Delta 9\,MACD$	Differences of technical index between the day and next day.
Moving average convergence and divergence (MACD)	9 MACD	MACD shows the difference between a fast and slow exponential moving average (EMA) of closing prices. Fast means a short-period average, and slow means a long period one.

Step 4. Calculate the partial F value of the original data for the technical index υ_j. If the F value is smaller than the user-defined threshold, υ_j is removed from the regression model since it does not affect the stock price significantly.

Step 5. Repeat Step 3 to Step 4. If the F value of variable is more than the user-defined threshold, the variable should be added to the model, otherwise it should be removed.

In addition, the range of the input variables of SVR model should be between 0 and 1. Hence, the selected technical indices are normalized as follows:

$$\text{Normal}(x_{ij}) = \frac{x_{ij} - \text{Min}(x_i)}{\text{Max}(x_i) - \text{Min}(x_i)} \quad i = 1,\ldots,n;\ j = 1,\ldots,m;\ n,\ m \in \Re,$$

(4.5)

where Normal (x_{ij}) denotes the normalized value of jth data point of ith technical index. Max (x_i) denotes the maximum value of ith technical index. Min (x_i) denotes the minimum value of ith technical index. x_{ij} denotes original value of jth data point of ith technical index. n and m denote the total number of technical indices and data points, respectively

Learning the Trading Forecasting Model by SVR

Support vector regression will be applied as a machine learning model to extract the hidden knowledge in the historic stock database. The single output is the trading signal ts from TSBM process, and the multiple input features are technical indices from SRA selection. SVR learning model transforms multiple features into high multidimensional feature space, and the transformed feature space can be mapped into a hyperplane space to determine correct signals based on those support vector points. On the kernel function selection, we try to use linear, RBF, polynomial, and sigmoid functions to generate better performance for the SVR model because the stock market is a very complicated nonlinear environment. Since the SVR approach possesses high learning capability and accuracy in predicting continuous signals for building hidden knowledge among trading signals and technical indices, it is a widely used tool for predicting the trading signals.

Trading Points Decision from Forecasted Trading Signal

In the daily forecasting, if the forecasted trading signals by SVR satisfied buy threshold, then this means it is needed to buy stock quickly because it is very close to turning point; otherwise if the state satisfied a sell threshold, then there is need to sell stock. These satisfied points are recommended to transaction in stock market. Before determining the trading point, we will calculate the buy/sell threshold values for two trading types. The trading thresholds of two types are as follows:

$$\text{Buy}_{\text{threshold}} = \mu + \sigma,$$

$$\text{Sell}_{\text{threshold}} = 1 - \mu + \sigma,$$

$$\mu = \frac{1}{N}\sum_{i=1}^{N} x_i',$$

$$\sigma = \sqrt{\frac{1}{N}\sum_{i=1}^{N}(x_i' - \mu)},$$

$$(4.6)$$

where μ denotes the average of trading signal in training data. σ denotes the standard deviation of trading signal in training data. Buy_ threshold denotes the buy trading threshold. Sell_threshold denotes the sell trading threshold. If forecasted trading signals form SVR model in testing data are more than buy_threshold, then this suggests trading point for buy stocks else if forecasting signal in testing data is smaller than sell_threshold, then this suggests trading for sell stock.

In the trading decision step, the TBSM-SVR model is employed to calculate daily trading signals. The detailed principles for making trading decisions include the following.

1. If the time series prediction of trading signals by TBSM-SVR model is going up and intersects with buy trading threshold Buy_threshold, then it is a "buy" trading decision.
2. If the time series prediction of trading signals by TBSM-SVR model is going down and intersects with sell trading threshold sell_threshold, then it is a "sell" trading decision.
3. A "hold" trading decision is made (or do not make any trading decision) when the forecasting trading signal does not intersect with buy and sell thresholds.

For example, Figure 5 shows trading points decision for Apple stock. How to suggest the buy/sell points for stock in a time series in which the red square points are buy points and green triangle points are the sell points? Both are satisfied two thresholds in which the orange dotted line is sell threshold and the purple dotted line is buy threshold, so we can forecast the trading points daily by an automatically trading system.

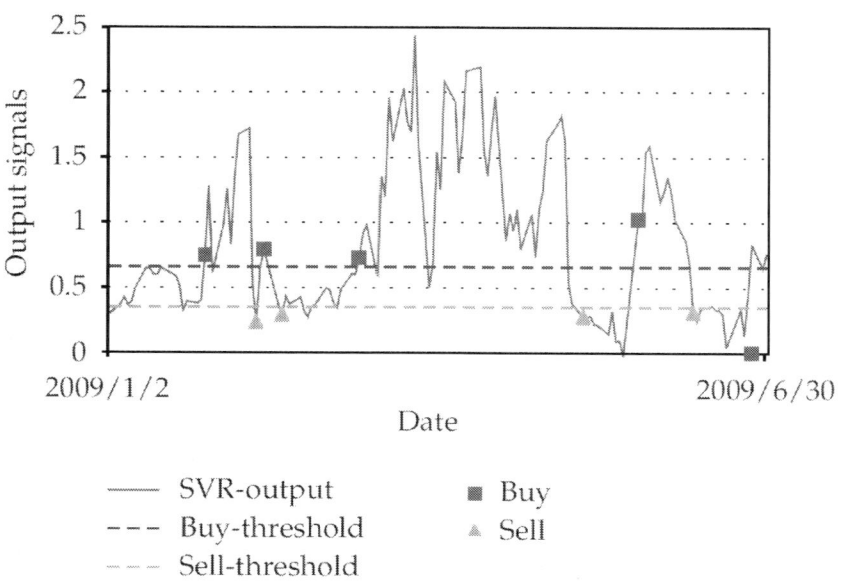

Figure 5: An example of result for detecting trading points of Apple.

EXPERIMENTAL RESULTS

Profit Evaluation and Parameters Setting

In this research, the trading point (buy and sell timing) is decided by the TBSM-SVR model based on the forecasting trading signal of SVR and TBSM segmentation. In the experimental section, we also use various forecasting models to the generated profiting trading points and compare their performances. The profits in each different forecasting model are calculated as follows:

$$\text{profits} = C \prod_{i=1}^{k} \left\{ \frac{(1 - a - b) \times p_{S_i} - (1 + a) \times p_{B_i}}{(1 + a) \times p_{B_i}} \right\}, \tag{5.1}$$

where C is the total amount of money to be invested at the beginning as well as the capital of money, a refers to the tax rate of ith transaction, b refers to the handling charge of ith transaction, k is the total number of transaction, p_{S_i} is the selling price of the ith transaction and p_{B_i} is the buying price of ith transaction.

This study uses minimal root mean square error (RMSE) to measure the model performance in SVR train stage. In the model selection strategy that the dataset uses the last one trading period of training data contains (buy/sell and sell/buy states). The RMSE of an estimator $\widehat{ts}$ with respect to the estimated parameter ts is defined as the square root of the mean square error:

$$\text{RMSE} = \sqrt{\frac{\sum_{i=1}^{n} ts_i - \widehat{ts_i}}{N}}. \tag{5.2}$$

ts denotes the trading signal by trading signal transformation from TBSM segmentation in Section 4.2. $\widehat{ts}$ denotes the estimated trading signal by SVR forecasting model. N denotes total number in each training data (Table 2).

Table 2: The parameter setup for TBSM and SVR by DOEs (design of experiments)

Approach	Parameter	Value	Explanation
TBSM	Threshold	0.1 σ to 1 σ	The difference of price at uptrend or downtrend
TBSM	X_Thld	0.1 σ to 1 σ	The difference of days at hold trend
TBSM	Y_Thld	0.1 σ to 1 σ	The difference of price at hold trend
SVR	C	^{3}to^3	Cost
SVR	ε	^{4}to	Epsilon
SVR	d	^{9}to	Degree
SVR	g	2^1to2^4	Gamma

In parameter section we use design of experiments (DOEs) approach to set each parameter for capture optimal parameter combination for trading system in financial data. The parameters of the TBSM are based on standard deviation σ from stock price in each stock which is the range from 0.1 σ to 1 σ for testing in each parameters. In SVR model, the kernels chosen for testing are "radial basis function (RBF)" and "polynomial" function. The common combination includes cost C;epsilon and are selected by the grid search with exponentially growing sequences. C ranges from 10^{-3} to 10^3. ε from 10^{-4} to 10^{-1} and γ is fixed as 0. In "polynomial" function, the degree d ranges from 2^{-9} to.2^{-1} The gamma g ranges from 2^1 to 2^4 in RBF kernel.

Profit Comparison in the US Stock Market

In this research, we have selected 7 stocks from the US stock market to compare the profit achieved by various trading models, including Apple, BOENING CO. (BA), Caterpillar Inc. (CAT), Johnson and Johnson (JNJ), Exxon Mobil Corp. (XOM), Verizon Communication Inc. (VZ), and S&P 500. Among all the stocks, 253 data points were collected for the training period from 1/2/2008 (mm/dd/yy) to 12/31/2008 while 124 data points were used for the testing period from 1/2/2009 to 6/30/2009. In this research, we have compared our forecasting model of TBMS-SVR approach with two other identification models developed in the past. The PLR-BPN model proposed by Chang et al. [26] used neural networks in combination with PLR and exponential smoothing to determine the trading points. Kwon and Kish [41] used statistical model such as moving average, rate of change and trading volumes to determine the buy-sell points and generated profit.

The technical incices selected result by SRA as shown in Table 3. Apple, Ba, CAT, JNJ, S&P 500, and VZ used 5 features (technical indices) for training forecasting model; XOM used 3 features for training forecasting model. From this result we can know that a few features can capture more trading knowledge.

Table 3: Feature selection result in each stock for technical indices by SRA

Stock	Technical index
Apple	5 MA, 6 MA, 9 K, 9 MACD, 12 W%R
BA	5 MA, 6 MA, 9 K, 10 TV, 12 W%R
CAT	5 MA, 6 MA, 9 K, 10 TV, 5 MA
JNJ	5 MA, 6 MA, 6 RSI, 9 MACD, 5 MA
S&P 500	5 MA, 5 BIAS, 10 TV, 26 BR, TAPI
VZ	5 MA, 6 MA, 5 MA, 10 TV, 26 VR
XOM	5 MA,6 MA, 5 MA

From model selection results the RBF kernel has better low error in each stock by RMSE. Moreover, the gamma, degree, cost, epsilon, support vectors, and RMSE as shown in Table 4 are necessary parameters and measures. The models of TBSM-SVR in each stock are selecting optimal parameter combination by RMSE consideration.

Table 4: Model selection results from TSBM-SVR model for each stock

Stock	Kernel									
	Radial basis function (RBF)			Polynomial						
	g	C	SVs	RMSE	d	C			SVs	RMSE
Apple	2^{-1}	10^3	10^{-4}	253	0.0819	2	[0.001 : 1000]	[0.0001 : 0.1]	71	0.266
BA	2^{-1}	10^3	10^{-1}	107	0.0955	2	[0.001 : 1000]	[0.0001 : 0.1]	76	0.269
CAT	2^{-1}	10^3	10^{-3}	254	0.0898	2	[0.001 : 1000]	[0.0001 : 0.1]	156	0.233
JNJ	2^{-1}	10^2	10^{-1}	137	0.2617	1	[0.001 : 1000]	[0.0001 : 0.1]	116	0.426
S&P 500	2^{-1}	10^3	10^{-4}	254	0.0004	1	[0.001 : 1000]	[0.0001 : 0.1]	112	0.379
VZ	2^{-1}	10^3	10^{-3}	251	0.0031	1	[0.001 : 1000]	[0.0001 : 0.1]	125	0.269
XOM	2^{-1}	10^3	10^{-4}	253	0.0001	2	[0.001 : 1000]	[0.0001 : 0.1]	182	0.18

Table 5: Comparison of profit obtained by various forecasting models

Stock no.	Stock name	TBSN-SVR model (RBF)	PLR-SVR model (RBF)	PLR-BPN model	Statistical model
1	Apple	92.35%	35.84%	12.97%	20.50%
2	BA	59.49%	35.69%	17.50%	20.03%
3	CAT	43.39%	36.09%	9.36%	24.83%
4	JNJ	13.95%	9.47%	16.88%	0%
5	S&P 500	22.78%	4.19%	3.77%	9.81%
6	VZ	28.60%	2.60%	27.72%	0%
7	XOM	22.40%	12.34%	−1.99%	−7.65%
Average		40.42%	19.46%	12.32%	9.65%

Each forecasting model provides trading points for each stock, so the best profits of the 3 forecasting models are shown in Table 5. The results turn out that our proposed TBSM with SVR model generates the greatest returns for the seven stocks, that is, number 1, 2, 3, 4, 5, 6, and 7 outperform other models. The average profit rate of these seven stocks is 40.42% using the TBSM-SVR model whereas the average profit rate generated by other models like PLR-SVR, PLR-BPN, and Statistical is 19.46%, 12.32%, and 9.65%, respectively. Therefore, our TBSM approach is better than PLR approach which is only considered linear representation.

The buy and sell points obtained from the TBSM forecasting model in each stock are shown in Figures 6, 7, 8,9, 10, 11, and 12. The red square represents the buy point, and the black triangle represents the sell point using a trading strategy to determine turning points. Furthermore, our proposed approach TBSM is better than PLR segmentation which denotes that TBSM approach captures better trading knowledge for SVR forecasting model. Due to PLR only the linear representation is considering, so it loses important trend. Therefore, TBSM is an effective segmentation method for nonlinear time series data in stock market.

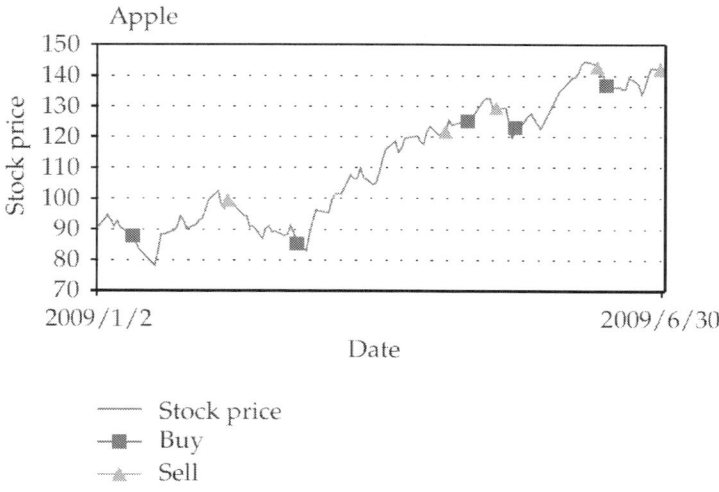

Figure 6: The forecasted trading points of Apple (an uptrend stock).

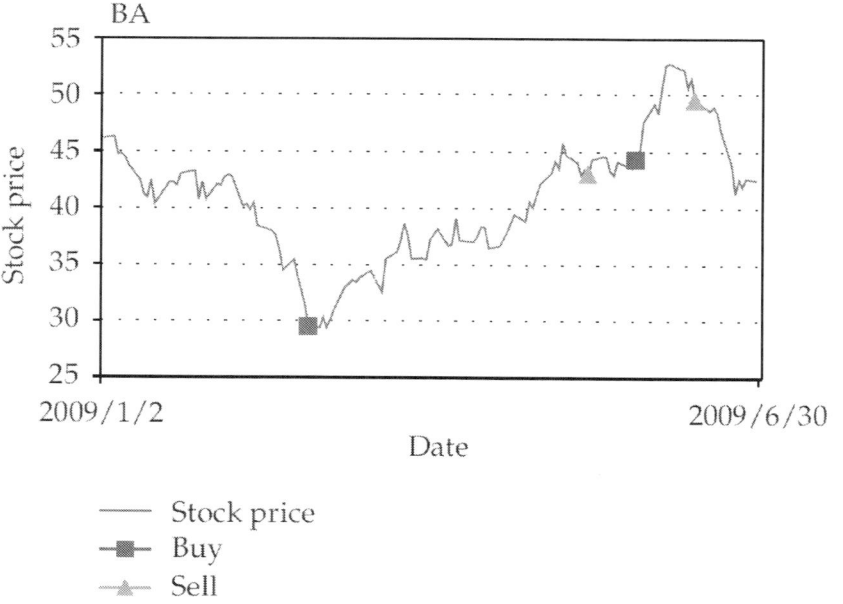

Figure 7: The forecasted trading points of BA (a steady-trend stock).

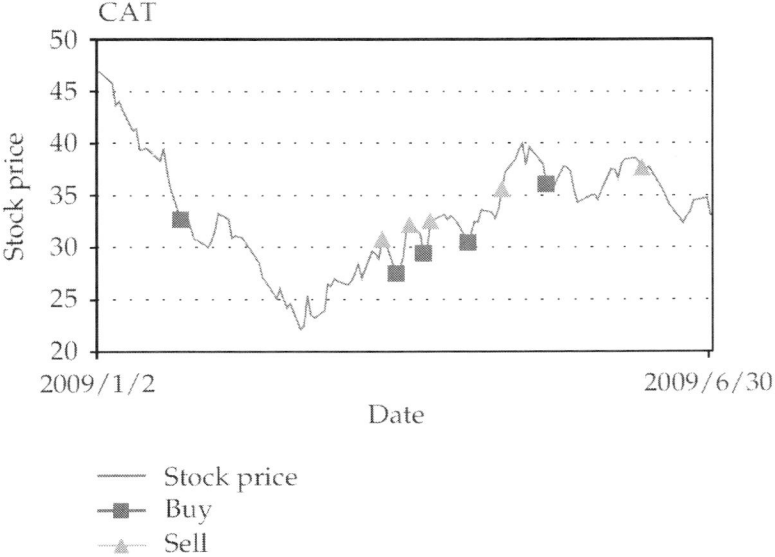

Figure 8: The forecasted trading points of CAT (a downtrend stock).

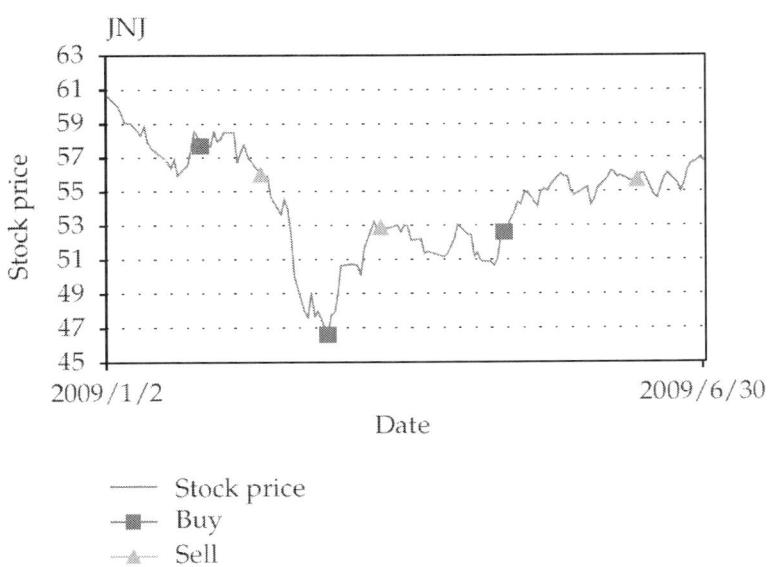

Figure 9: The forecasted trading points of JNJ (a steady-trend stock).

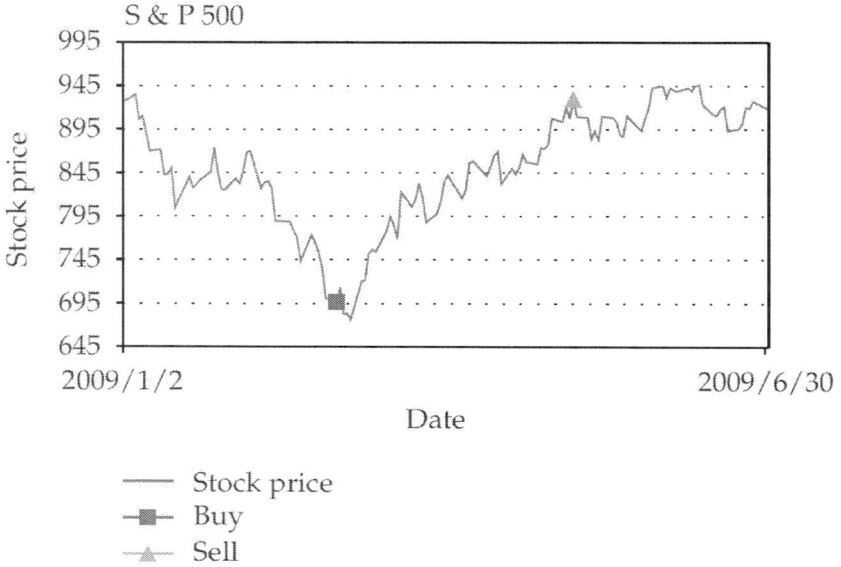

Figure 10: The forecasted trading points of S&P 500 (a steady-trend stock).

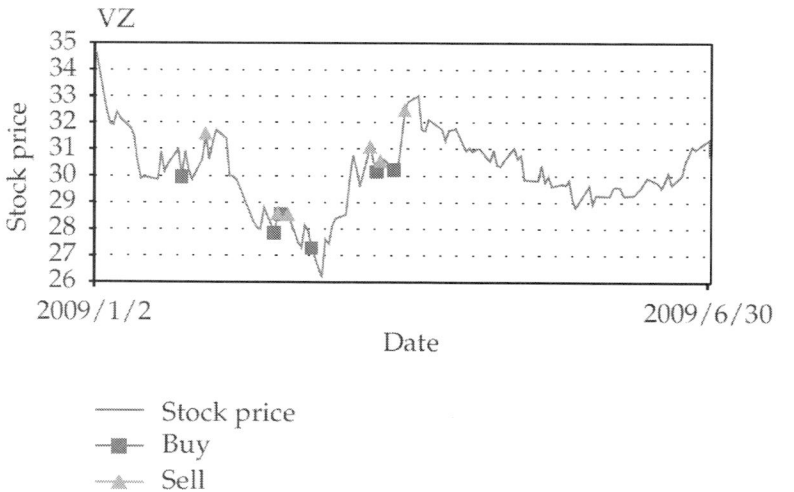

Figure 11: The forecasted trading points of VZ (a downtrend stock).

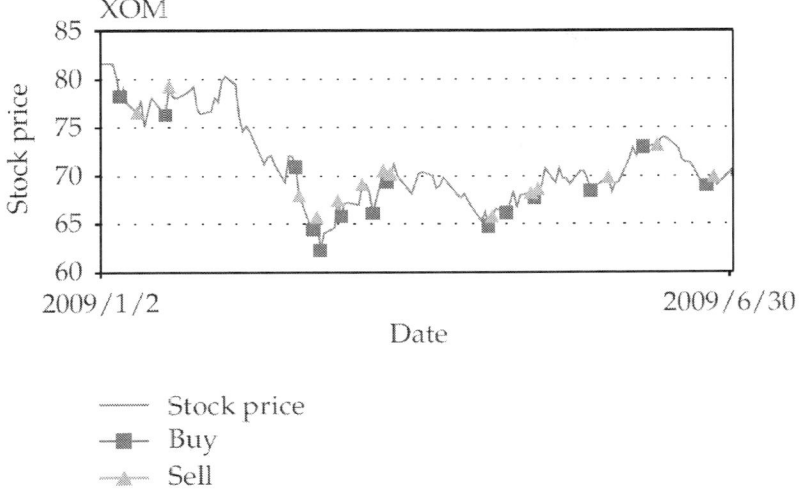

Figure 12: The forecasted trading points of XOM (a downtrend stock).

CONCLUSIONS

In this paper we proposed a trading system combining TBSM with SVR, and it is called TBSM-SVR-based stock trading system. This new trading system has been very effective in earning high profit while with the greatest ability. Experimental results showed that the TSBM can segment the stock price's variation into different trading trends. The trading signal in each trading trend will be assumed to be the same. The nonlinear time series can be better represented using these trading trends. Additionally, SVR is applied to capture the trading knowledge using the trading signals derived from these trading trends. The captured knowledge is more effective using TBSM-SVR when compared to PLR segmentation method. As a result, the primary goal of the investor could be easily achieved by providing him with simple trading decisions. However, the limitation of the TBSM-SVR trading system is the machine learning tool; that is, SVR is still not that mature yet. There are still rooms for the improvement of a better machine learning mechanism to be developed. Therefore, the trading system may make a wrong trading and lose money. In the future works, we can extend the segmentation method by considering a more detailed trend by investigating different buy-hold strategy or better trading strategy. In addition, the trend based segmentation method can further consider the fractal properties such as long memory, which can be accommodated to improve the segmentation performances.

REFERENCES

1. V. N. Vapnik, The Nature of Statistical Learning Theory, Springer, New York, NY, USA, 1995.
2. V. N. Vapnik, Statistical Learning Theory, Adaptive and Learning Systems for Signal Processing, Communications, and Control, John Wiley & Sons, New York, NY, USA, 1998.
3. Z. Liu, "Chaotic time series analysis," Mathematical Problems in Engineering, vol. 2010, Article ID 720190, 31 pages, 2010. · ·
4. B. J. Chen, M. W. Chang, and C. J. Lin, "Load forecasting using support vector machines: a study on EUNITE Competition 2001," IEEE Transactions on Power Systems, vol. 19, no. 4, pp. 1821–1830, 2004. · ·
5. F. Girosi, M. Jones, and T. Poggio, "Regularization theory and neural networks architectures," Neural Computation, vol. 7, pp. 219–269, 1995.
6. X. H. Yang, D. X. She, Z. F. Yang, Q. H. Tang, and J. Q. Li, "Chaotic bayesian method based on multiple criteria decision making (MCDM) for forecasting non-linear hydrological time series," International Journal of Nonlinear Sciences and Numerical Simulation, vol. 10, no. 11-12, pp. 1595–1610, 2009. ·
7. D. She and X. Yang, "A new adaptive local linear prediction method and its application in hydrological time series," Mathematical Problems in Engineering, vol. 2010, Article ID 205438, 15 pages, 2010. · ·
8. N. Muttil and K. W. Chau, "Neural network and genetic programming for modelling coastal algal blooms," International Journal of Environment and Pollution, vol. 28, no. 3-4, pp. 223–238, 2006. · ·
9. D. Niu, Y. Wang, and D. D. Wu, "Power load forecasting using support vector machine and ant colony optimization," Expert Systems with Applications, vol. 37, no. 3, pp. 2531–2539, 2010. · ·
10. P. F. Pai and W. C. Hong, "Forecasting regional electricity load based on recurrent support vector machines with genetic algorithms," Electric Power Systems Research, vol. 74, no. 3, pp. 417–425, 2005. · ·

11. W. C. Hong, "Chaotic particle swarm optimization algorithm in a support vector regression electric load forecasting model," Energy Conversion and Management, vol. 50, no. 1, pp. 105–117, 2009. · ·

12. T. Farooq, A. Guergachi, and S. Krishnan, "Knowledge-based Green›s Kernel for support vector regression," Mathematical Problems in Engineering, vol. 2010, Article ID 378652, 16 pages, 2010. · ·

13. S. O. Lozza, E. Angelelli, and A. Bianchi, "Financial applications of bivariate Markov processes,"Mathematical Problems in Engineering, vol. 2011, Article ID 347604, 15 pages, 2011. ·

14. A. Swishchuk and R. Manca, "Modeling and pricing of variance and volatility swaps for local semi-markov volatilities in financial engineering," Mathematical Problems in Engineering, vol. 2010, Article ID 537571, 17 pages, 2010. · · ·

15. M. S. Abd-Elouahab, N. E. Hamri, and J. Wang, "Chaos control of a fractional-order financial system,"Mathematical Problems in Engineering, vol. 2010, Article ID 270646, 18 pages, 2010. · · ·

16. A. J. Smola and B. Schölkopf, "A tutorial on support vector regression," Statistics and Computing, vol. 14, no. 3, pp. 199–222, 2004. · ·

17. P. C. Chang and C. H. Liu, "A TSK type fuzzy rule based system for stock price prediction," Expert Systems with Applications, vol. 34, no. 1, pp. 135–144, 2008.

18. P. F. Pai and C. S. Lin, "A hybrid ARIMA and support vector machines model in stock price forecasting," Omega, vol. 33, no. 6, pp. 497–505, 2005. · ·

19. F. X. Diebold and R. S. Mariano, "Comparing predictive accuracy," Journal of Business and Economic Statistics, vol. 20, no. 1, pp. 134–144, 2002. · ·

20. H. Liu and J. Wang, "Integrating independent component analysis and principal component analysis with neural network to predict Chinese stock market," Mathematical Problems in Engineering, vol. 2011, Article ID 382659, 15 pages, 2011.

21. X. P. Ge, "Pattern matching in financial time series data," Computer Communications, vol. 27, pp. 935–945, 1998.

22. E. Keogh and M. Pazzani, "An enhanced representation of time series which allows fast and accurate classification, clustering and relevance feedback," in Proceedings of the 4th International Conference on Knowledge Discovery and Data Mining (KDD ‹98), pp. 239–241, August 1998.

23. V. Lavrenko, M. Schmill, D. Lawrie, P. Ogilvie, D. Jensen, and J. Allan, "Mining of concurrent text and time series," in Proceedings of the 6th International Conference on Knowledge Discovery and Data Mining (KDD ‹00), pp. 37–44, August 2000.

24. S. Ghosh, P. Manimaran, and P. K. Panigrahi, "Characterizing multi-scale self-similar behavior and non-statistical properties of fluctuations in financial time series," Physica A, vol. 390, no. 23-24, pp. 4304–4316, 2011. · ·

25. P. C. Chang, C. Y. Tsai, C. H. Huang, and C. Y. Fan, "Application of a case base reasoning based support vector machine for financial time series data forecast-

ing," in Proceedings of the International Conference on Intelligent Computing (ICIC ‹09), vol. 5755, pp. 294–304, September 2009.

26. P. C. Chang, C. Y. Fan, and C. H. Liu, "Integrating a piecewise linear representation method and a neural network model for stock trading points prediction," IEEE Transactions on Systems, Man and Cybernetics Part C, vol. 39, no. 1, pp. 80–92, 2009. · ·

27. L. Todorova and B. Vogt, "Power law distribution in high frequency financial data? An econometric analysis," Physica A, vol. 390, no. 23-24, pp. 4433–4444, 2011. ·

28. M. K. P. So, C. W. S. Chen, J. Y. Lee, and Y. P. Chang, "An empirical evaluation of fat-tailed distributions in modeling financial time series," Mathematics and Computers in Simulation, vol. 77, no. 1, pp. 96–108, 2008. · ·

29. M. Li and W. Zhao, "Visiting power laws in cyber-physical networking systems," Mathematical Problems in Engineering, vol. 2012, Article ID 302786, 13 pages, 2012.

30. L. Muchnik, A. Bunde, and S. Havlin, "Long term memory in extreme returns of financial time series,"Physica A, vol. 388, no. 19, pp. 4145–4150, 2009. · ·

31. M. Li, C. Cattani, and S. Y. Chen, "Viewing sea level by a one-dimensional random function with long memory," Mathematical Problems in Engineering, vol. 2011, Article ID 654284, 13 pages, 2011. · ·

32. M. Li, "Fractal time series—a tutorial review," Mathematical Problems in Engineering, vol. 2010, Article ID 157264, 26 pages, 2010. · ·

33. J. O. Lachaud, A. Vialard, and F. De Vieilleville, "Analysis and comparative evaluation of discrete tangent estimators," in Proceedings of the 12th International Conference on Discrete Geometry for Computer Imagery (DGCI ‹05), E. Andres, G. Damiand, and P. Lienhardt, Eds., vol. 3429,, pp. 240–251, Springer, April 2005.

34. Y. Zhu, D. Wu, and S. Li, "A piecewise linear representation method of time series based on feature pints," in Proceedings of the 11th International Conference on Knowledge-Based Intelligent Information and Engineering Systems (KES ‹07), 17th Italian Workshop on Neural Networks (WIRN ‹07), pp. 1066–1072, January 2007.

35. H. Wu, B. Salzberg, and D. Zhang, "Online event-driven subsequence matching over financial data streams," in Proceedings of the ACM SIGMOD International Conference on Management of Data (SIGMOD ‹04), pp. 23–34, June 2004.

36. Z. Zhang, J. Jiang, X. Liu et al., "Pattern recognition in stock data based on a new segmentation algorithm," in Proceedings of the 2nd International Conference on Knowledge Science, Engineering and Management (KSEM ‹07), vol. 4798 of Lecture Notes in Computer Science, pp. 520–525, 2007.

37. Y. W. Wang, P. C. Chang, C. Y. Fan, and C. H. Huang, "Database classification by integrating a case-based reasoning and support vector machine for induction," Journal of Circuits, Systems and Computers, vol. 19, no. 1, pp. 31–44, 2010. · ·

38. L. Zhang, W. D. Zhou, and P. C. Chang, "Generalized nonlinear discriminant analysis and its small sample size problems," Neurocomputing, vol. 74, no. 4, pp. 568–574, 2011. · ·
39. N. Sapankevych and R. Sankar, "Time series prediction using support vector machines: a survey," IEEE Computational Intelligence Magazine, vol. 4, no. 2, pp. 24–38, 2009. · ·
40. J. L. Wu, L. C. Yu, and P. C. Chang, "Emotion classification by removal of the overlap from incremental association language features," Journal of the Chinese Institute of Engineers, vol. 34, no. 7, pp. 947–955, 2011.
41. K. Y. Kwon and R. J. Kish, "Technical trading strategies and return predictability: NYSE," Applied Financial Economics, vol. 12, no. 9, pp. 639–653, 2002. · ·

CITATION

Jheng-Long Wu and Pei-Chann Chang, "A Trend-Based Segmentation Method and the Support Vector Regression for Financial Time Series Forecasting," Mathematical Problems in Engineering, vol. 2012, Article ID 615152, 20 pages, 2012. doi:10.1155/2012/615152.

Index